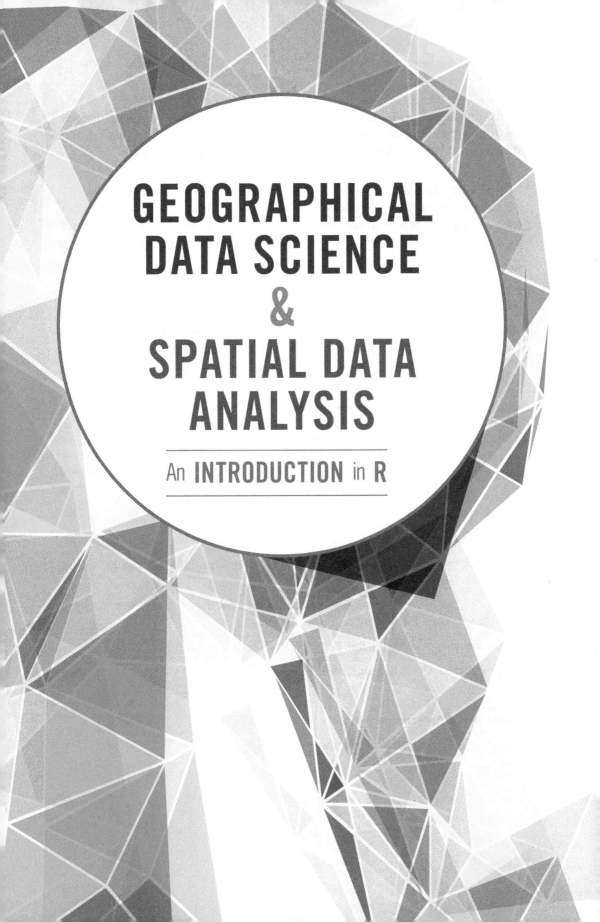

GEOGRAPHICAL DATA SCIENCE
&
SPATIAL DATA ANALYSIS

An **INTRODUCTION** in R

In the digital age, social and environmental scientists have more spatial data at their fingertips than ever before. But how do we capture this data, analyse and display it, and most importantly, how can it be used to study the world?

Spatial Analytics and GIS is a series of books that deal with potentially tricky technical content in a way that is accessible, usable and useful. Titles include *Urban Analytics* by Alex Singleton, Seth Spielman and David Folch, and *An Introduction to R for Spatial Analysis and Mapping (Second Edition)* by Chris Brunsdon and Lex Comber.

Series Editor: Richard Harris

About the Series Editor

Richard Harris is Professor of Quantitative Social Geography at the School of Geographical Sciences, University of Bristol. He is the lead author on three textbooks about quantitative methods in geography and related disciplines, including *Quantitative Geography: The Basics* (Sage, 2016).

Richard's interests are in the geographies of education and the education of geographers. He is currently Director of the University of Bristol Q-Step Centre, part of a multimillion pound UK initiative to raise quantitative skills training among social science students, and is working with the Royal Geographical Society (with IBG) to support data skills in schools.

Books in this Series:

Geographical Data Science and Spatial Data Analysis, Lex Comber & Chris Brunsdon

An Introduction to R for Spatial Analysis and Mapping, 2nd Edition, Chris Brunsdon & Lex Comber

Agent-Based Modelling and Geographical Information Systems, Andrew Crooks, Nicolas Malleson, Ed Manley & Alison Heppenstall

Urban Analytics, Alex Singleton, Seth Spielman & David Folch

Geocomputation, Chris Brunsdon & Alex Singleton

Published in Association with this Series:

Quantitative Geography, Richard Harris

GEOGRAPHICAL DATA SCIENCE

&

SPATIAL DATA ANALYSIS

An **INTRODUCTION** in R

LEX COMBER

and

CHRIS BRUNSDON

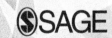

Los Angeles | London | New Delhi
Singapore | Washington DC | Melbourne

Los Angeles | London | New Delhi
Singapore | Washington DC | Melbourne

SAGE Publications Ltd
1 Oliver's Yard
55 City Road
London EC1Y 1SP

SAGE Publications Inc.
2455 Teller Road
Thousand Oaks, California 91320

SAGE Publications India Pvt Ltd
B 1/I 1 Mohan Cooperative Industrial Area
Mathura Road
New Delhi 110 044

SAGE Publications Asia-Pacific Pte Ltd
3 Church Street
#10-04 Samsung Hub
Singapore 049483

Editor: Jai Seaman
Assistant editor: Charlotte Bush
Assistant editor, digital: Sunita Patel
Production editor: Katherine Haw
Copyeditor: Richard Leigh
Proofreader: Neville Hankins
Indexer: Martin Hargreaves
Marketing manager: Susheel Gokarakonda
Cover design: Francis Kenney
Typeset by: C&M Digitals (P) Ltd, Chennai, India

Library of Congress Control Number: 2020938055

British Library Cataloguing in Publication data

A catalogue record for this book is available from the British Library

ISBN 978-1-5264-4935-1
ISBN 978-1-5264-4936-8 (pbk)

Lex: To my children, Carmen, Fergus and Madeleine: you are all adults now. May you continue to express your ever growing independence in body, as well as thought.

Chris: To all of my family – living and no longer living.

CONTENTS

ABOUT THE AUTHORS

Alexis Comber (Lex) is Professor of Spatial Data Analytics at Leeds Institute for Data Analytics (LIDA), University of Leeds. He worked previously at the University of Leicester where he held a chair in Geographical Information Sciences. His first degree was in Plant and Crop Science at the University of Nottingham and he completed a PhD in Computer Science at the Macaulay Institute, Aberdeen (now the James Hutton Institute), and the University of Aberdeen. This developed expert systems for land cover monitoring from satellite imagery and brought him into the world of spatial data, spatial analysis and mapping. Lex's research interests span many different application areas including environment, land cover/land use, demographics, public health, agriculture, bio-energy and accessibility, all of which require multi-disciplinary approaches. His research draws from methods in geocomputation, mathematics, statistics and computer science, and he has extended techniques in operations research/location allocation (what to put where), graph theory (cluster detection in networks), heuristic searches (how to move intelligently through highly dimensional big data), remote sensing (novel approaches for classification), handling divergent data semantics (uncertainty handling, ontologies, text mining) and spatial statistics (quantifying spatial and temporal process heterogeneity). He has co-authored (with Chris Brunsdon) the first 'how to' book for spatial analysis and mapping in R, the open source statistical software, now in its second edition (https://uk.sagepub.com/en-gb/eur/an-introduction-to-r-for-spatial-analysis-and-mapping/book258267). Outside of academic work and in no particular order, Lex enjoys his vegetable garden, walking the dog and playing pinball (he is the proud owner of a 1981 Bally Eight Ball Deluxe).

Chris Brunsdon is Professor of Geocomputation and Director of the National Centre for Geocomputation at the National University of Ireland, Maynooth, having worked previously in the Universities of Newcastle, Glamorgan, Leicester and Liverpool, variously in departments focusing on both geography and computing. He has interests that span both of these disciplines, including spatial statistics, geographical information science, and exploratory spatial data analysis, and in particular the application of these ideas to crime pattern analysis, the modelling of house prices, medical and health geography and the analysis of land use data. He was one of the originators of the technique of geographically weighted regression (GWR). He has extensive experience of programming in R, going back to the late

1990s, and has developed a number of R packages which are currently available on CRAN, the Comprehensive R Archive Network. He is an advocate of free and open source software, and in particular the use of reproducible research methods, and has contributed to a large number of workshops on the use of R and of GWR in a number of countries, including the UK, Ireland, Japan, Canada, the USA, the Czech Republic and Australia. When not involved in academic work he enjoys running, collecting clocks and watches, and cooking – the last of these probably cancelling out the benefits of the first.

PREFACE

Data and data science are emerging (or have emerged) as a dominant activity in many disciplines that now recognise the need for empirical evidence to support decision-making (although at the time of writing in the UK at the end of April 2020, this is not obvious). All data are spatial data – they are collected some*where* – and location cannot be treated as just another variable in most statistical models. And because of the ever growing volumes of (spatial) data, from increasingly diverse sources, describing all kinds of phenomena and processes, being able to develop approaches and models grounded in *spatial* data analytics is increasingly important. This books pulls together and links lessons from general data science to those from quantitative geography, which have been developed and applied over many years.

In fact, the practices and methods of data science, if framed as being a more recent term for statistical analysis, and spatial data science, viewed as being grounded in geographical information systems and science, are far from new. A review of the developments in these fields would suggest that the ideas of data analytics have arisen as a gradual evolution. One interesting facet of this domain is the importance of spatial considerations, particularly in marketing, where handling locational data has been a long-standing core activity. The result is that geographical information scientists and quantitative geographers are now leading many data science activities – consider the background of key players at the Alan Turing Institute, for example. Leadership is needed from this group in order to ensure that lessons learned and experiences gained are shared and disseminated. A typical example of this is the *modifiable areal unit problem,* which in brief posits that statistical distributions, relationships and trends exhibit very different properties when the same data are aggregated or combined over different areal units, and at different spatial scales. It describes the process of distortion in calculations and differences in outcomes caused by changes zoning and scales. This applies to **all** analyses of spatial data – and has universal consequences, but is typically unacknowledged by research in non-geographical domains using spatial data.

This possibility of distortion also underpins another motivation for this book at this time: one of *reproducibility.* The background to R, the open source statistical package, is well documented and a number of resources have been published that cover recent developments in the context of spatial data and spatial analysis in R (including our other offering in this arena: Brunsdon and Comber, 2018). This has promoted the notion of the need for *open* coding environments within which analysis takes place, thereby allowing (spatial) data science cultures to flourish.

And in turn this has resulted in a de facto way of working that embraces open thinking, open working, sharing, open collaboration, and, ultimately, reproducibility and transparency in research and analysis. This has been massively supported by the RStudio integrated development environment for working in R, particularly the inclusion of RMarkdown which allows users to embed code, analysis and data within a single document, as well as the author's interpretation of the results. This is truly the 'holy grail' of scientific publishing!

A further driver for writing this book is to promote notions of *critical data science*. Through the various examples and illustration in the book, we have sought to show how different answers/results (and therefore understandings and predictions) can be generated by very small and subtle changes to models, either through the selection of the machine learning algorithm, the scale of the data used or the choice of the input variables. Thus we reject *plug and play* data science, we reject the idea of theory-free analyses, we reject data mining, all of which abrogate inferential responsibility through philosophies grounded in *letting the data speak*. Many of the new forms of data that are increasingly available to the analyst are not objective (this is especially the case for what has sometimes been called 'big data'). They are often collected without any experimental design, have many inherent biases and omissions, and without careful consideration can result in erroneous inference and poor decision-making. Thus being *critical* means considering the technological, social and economic origins of data, including their creation and deployment, as well as the properties of the data relative to the intended analysis, or the consequences of any analysis. Criticality involves thinking about the common good, social contexts, using data responsibly, and even considering how your work could be used in the wrong way or the results misinterpreted. There is no excuse for number crunchers who fail to be critical in their data analysis.

In summary, we believe that the practice of data analytics (actually *spatial* data analytics) should be done in an open and reproducible way, it should include a critical approach to the broader issues surrounding the data, their analysis and consideration of how they will be used, and it should be done wearing geography goggles to highlight the impacts of scale and zonation on the results of analyses of spatial data. This may involve some detective work to understand the impacts of data and analysis choices on the findings – this too is a part of data science. We believe that the contents of this book, and the various coded examples, provide the reader with an implicit grounding in these issues.

REFERENCE

Brunsdon, C. and Comber, L. (2018) *An Introduction to R for Spatial Analysis and Mapping* (2nd edn). London: Sage.

ONLINE RESOURCES

Geographical Data Science and Spatial Data Analysis: An Introduction in R is supported by online resources. Find them at: https://study.sagepub.com/comber.

Get confident using R with a **code library** of up-to-date R scripts from each chapter.

Practice your skills on real-world data, with datasets from the **data library**.

Deepen your understanding with a set of curated **journal articles** on important topics such as critical spatial data science.

1

INTRODUCTION TO GEOGRAPHICAL DATA SCIENCE AND SPATIAL DATA ANALYTICS

1.1 OVERVIEW

The aim of this book is to provide an introduction to both data and spatial data analytics. These encompass data manipulation, exploratory data analysis, visualisation, modelling, creating and integrating databases, hypothesis testing and data analysis, and the communication of results via sophisticated multi-level graphics. The unique selling point of this book is that it extends activities in data science to the spatial case. An explicit treatment of spatial data is needed because increasingly data are spatial (as described below), the outputs of data analyses are increasingly mapped and the analysis of spatial data involves some particular considerations.

The book describes a framework for integrated and critical (spatial) data science. It draws extensively from the integrated suite of tools and packages that are wrapped up in the `tidyverse` package (Wickham et al., 2019). This promotes consistent and tidy data formats and data manipulations (https://www.tidyverse.org). Such frameworks also link analytical methods to remote databases, avoiding the need for *in-memory* data, for exploratory data analysis and visualisations, and for working with spatial data. Throughout the book we emphasise the importance of considering the spatial properties of data, supported by integrated frameworks for spatial data such as the `sf` package (Pebesma et al., 2019). Additionally we have tried to remove some of the hype around machine learning, and instead stressed the need for clear distinctions between analytical objectives, data and their potential limitations and the approaches used to achieve those objectives, for example between inference (process understanding) and prediction.

The book contents describe a sequence of activities that move the reader through data and spatial data, visualisations, databases, machine learning and alternative spatial representations. It reflects what we know now, and inevitably this changes almost every day as activity and development in this area continue apace. However, for both of us, the most salient aspects have been the inherent and necessary critical reflection on the many misplaced assumptions around data science. To this end, we have sought to implicitly promote a critical approach to data science, and indeed the process of writing this book has stimulated our thinking around reproducibility (Brunsdon and Comber, 2020b) and the need for criticality when working with very large datasets (Brunsdon and Comber, 2020a). We encourage readers to challenge the notion of data science objectivity, to adopt critical approaches (Kitchin and Lauriault, 2014), and to approach data science with an eye on the common good and social context of analysis. Most importantly, we emphasise the need for an awareness of the impacts of outcomes of such activities, reflecting O'Neil and Schutt (2014, p. 354) who state that 'even if you are honestly skeptical of your model, there is always the chance that it will be used the wrong way in spite of your warnings'.

This introductory chapter has four further sections. Section 1.2 sets the scene and describes the motivation for the book. It describes how the contents of each chapter are linked and provides information to potential tutors about learning arcs, indicating how the book components could be used to construct potential modules. Sections 1.3 and 1.4 provide an introduction to R for the inexperienced reader. Section 1.3 describes how to install R and RStudio, the concept of packages, and emphasises key ways of working in R (using scripts, setting up workspaces, etc.). Section 1.4 provides an introduction to assignment in R, and describes some basic object types in R as well some links to other introductory materials. Section 1.5 provides a summary of the chapter.

1.2 ABOUT THIS BOOK

1.2.1 Why *Geographical Data Science and Spatial Data Analytics?*

The focus of this book is very much on methods for data analytics and, more correctly, *spatial* data analytics. It emphasises the importance of explicitly considering the *spatial* properties of data through worked examples. We believe that this is important for a number of reasons.

The first reason is the increasing amount of digital data describing an increasing variety of activities. The digital footprints that we generate as part of our everyday lives are growing constantly, as nearly all of our transactions in life generate electronic data in some form: from financial transactions to leisure activities as well as credit, medical and social security records and flows of people through transportation networks. The age of so called *big data* (Kitchin, 2013; Kitchin and McArdle, 2016) is upon us and many authors have (alliteratively) characterised this: from the three, five or seven Vs (Laney, 2001; Marr, 2014; McNulty, 2014) to the three Ds of dynamic, diverse, dense – to which dirty could be added (Comber et al., 2016). The size of big data, and it is just 'data' now, makes it increasingly difficult to handle using traditional data structures (e.g. Excel) and traditional analysis techniques. Data are noisy, messy, unrepresentative, biased and mis-sampled. Because of these properties, many current datasets present a challenge to standard ways of data handling: they will not fit in an Excel spreadsheet, standard statistics have problems with inferences (everything becomes significant), and they challenge capacity (i.e. computers and software may not have enough memory to store, process and handle them). Together these properties suggest a fundamental breakage to traditional methods of data analysis, which are underpinned by collecting as little data as possible, under an experimental design (e.g. Myers et al., 2013), in order to ensure that any statistical inferences from data analysis are robust.

The second reason is the increasingly *spatial* nature of much digital data. Much of the data we generate has some form of locational attribute attached to it, either directly in the form of latitude and longitude or projected coordinates (easting and northing), or indirectly through the transaction address, postcode, geographic area (e.g. related to census, political or health geographies), or some other unique identifier that can be linked to location. The inclusion of location with data has been driven by a range of factors, not least of which is the pervasiveness of GPS- and web-enabled devices (e.g. smartphones, tablets and other devices) through which people collect and share information via social media, blogs and smartphone apps. Thus it possible to argue that all data are collected some*where*. The inclusion of some form of *geography* with data confers a critical advantage: the locational attributes (coordinates) can be used directly as input variables to models themselves or indirectly via their spatial frameworks. These can be used to link to other data by undertaking some sort of spatial intersection with, for example, data reported in census areas.

A further issue is that, statistically, location (easting and northing, or latitude and longitude) cannot be treated as just another variable. Many environmental, socio-economic and physical processes exhibit some form of spatial pattern which does not exhibit random or normal distributions: data observations may not be independent of their neighbours. Rather, nearby values are frequently spatially autocorrelated (clustered), resulting in hotspots and coldspots, with geographical trends, and the relationship between variables may be non-stationary – they may vary across geographical space. Consideration of such spatial heterogeneity and the spatial autocorrelation of values reflects Tobler's First Law of Geography (Tobler, 1970) which states that: 'Everything is related to everything else, but near things are more related to each other.' For this reason many standard statistical approaches may be inadequate for spatial data analysis as they do not account for the spatial interactions and complexities of processes described in the spatial data.

A final consideration is that spatial data may be subject to the ecological fallacy or modifiable areal unit problem, commonly known as the MAUP (Robinson, 1950; Openshaw, 1984a, 1984b). In outline, this posits that relationships, trends and spatial patterns may have different properties when data are aggregated or brought together over different scales of aggregation, confounding both prediction and inference. It is in the context of these developments in *spatial* data that this book has been written. It provides methods for dealing with these issues, for including and analysing data spatial properties and relationships that vary over space.

1.2.2 Why R?

Over the last few years, R and its integrated development environment RStudio have become the de facto tool for data science. For the novice user, R provides a great introduction to coding. It is relatively easy to learn and is very rewarding, especially early in the learning cycle, because of the tools available in contributed packages (or libraries), extending the core functionality of R. These contain sets of tools, functions and graphical routines for data analysis and visualisation, and are written and shared by expert R users. One example is the tidyverse package which brings together different tools (and packages) for data wrangling, integration, analysis and visualisation. It facilitates the sharing of skills, techniques, knowledge and expertise through open repositories such as GitHub (https://github.com) and RPubs (https://rpubs.com), where users share and publish code. These can be used to create documents, webpages, dashboards and mapping applications.

R has a number of other advantages over general-purpose programming languages such as Python that are related to:

- the core statistical functionality in R, which means that complex statistical models can be created with just a few lines of code (R was originally built for and by statisticians);
- its extensive choice of options for different and imaginative data visualisations;
- its flexibility – the same functionality can be achieved in many ways;
- its fertile ecosystem, in which leading-edge methods that are simply unavailable in commercial software are developed and shared by leading-edge researchers.

R was developed from the S language which was originally conceived at the Lucent Technologies (formerly AT&T) Bell Laboratories in the 1970s and 1980s. It was initially developed by Robert Gentleman and Ross Ihaka of the Department of Statistics at the University of Auckland. It has become widely used in many areas of scientific activity and quantitative research, partly because it is available for free but also because of its extensive and diverse functionality. This is through the continual growth in contributed code and functions in the form of R packages, which when installed can be called as libraries. The background to R, along with documentation and information about packages as well as the contributors, can be found at the R Project website (http://www.r-project.org). The last few years have seen an explosion of R packages and applications and the use of R as an analytical tool across a range of different scientific disciplines, and particularly in data science and data analytics. R provides an environment that allows many different data structures, analyses, tools and functions to be integrated according to the task in hand – a sort of data analysis Meccano (https://en.wikipedia.org/wiki/Meccano). Users are able to assemble the tools that *they* need for their particular analysis: to collect data via application programming interfaces (APIs), to pre-process and transform them, to analyse and visualise them in an almost infinite number of ways. The ability for do-it-yourself data collation, analysis and visualisation is reflected in the increased activity in these domains, and many textbooks introducing 'data science in R' have been published.

However, none of these explicitly address the spatial dimension of much data. Rather, any spatial properties are ignored in analysis and simply used to create maps. As indicated above, the locational and geographical properties of data require a slightly different treatment than standard analytical approaches: space *is* special. For these reasons, what we have sought to do is to write a user-friendly book with an explicitly geographical focus that reflects the latest developments in spatial data analytics in R. As you work through this book you will learn a number of techniques for using R directly to carry out spatial data analysis, visualisation and manipulation.

1.2.3 Chapter contents

The emergence of data analytics is an exciting development but needs to be undertaken within a critical context, with an understanding of how to generate robust inference and an awareness of the practical problems of working with spatial data. Specifically we argue against data fishing expeditions (so-called databating; Comber et al., 2016), preferring extensive detective work to develop theories and questions to be addressed through data analysis.

The book has nine chapters, and these should be read in sequence: each chapter builds on the previous one and provides core materials and concepts for the next.

In this chapter, Sections 1.3 and 1.4 describe ways of working in R, how to obtain R and RStudio, and provide some introductory materials for those who have never used R before. The chapter also describes generic but very useful procedures and guiding principles that we think are critical to successful data analysis as well as links to other materials.

Chapter 2 introduces *data* and *spatial data*, covering recent R developments that have resulted in new data formats. Key packages include `tidyverse`, a collection of packages for data analysis, and `sf` for spatial data. It introduces a number of visualisations, some mapping with the `tmap` package, and starts to hint at how a piping syntax can support efficient workflows. These can be difficult to understand at first, but they provide an incredibly efficient way to chain data operations together. The chapter also introduces different ways to manipulate data using the dplyr package which provides core 'data wrangling' functionality, allowing the user to prepare, join, filter and summarise data prior to analysis and visualisation.

Chapter 3 describes the underlying philosophies and approaches to data analysis in the tidyverse package and its components such as dplyr. It provides a formal and thorough treatment of piping syntax, demonstrating its application through tractable 'small' data examples, providing a clear and important bridge to Chapter 4.

Chapter 4 shows how R can be used to create and interrogate databases. Databases are core to data and spatial data analytics: large datasets in conventional flat data table formats (i.e. with lots of rows and columns) are problematic for many reasons, not least of which is the requirement to hold them in computer working memory in order to manipulate them. This chapter describes how to create remote databases, to populate them with large data tables, and how to query them using dplyr piping syntax. In this framework, only query results are returned to R's working memory, providing a practical underpinning analysis of very large datasets. The chapter includes a worked example analysing 120 million medical prescription records held on a remote database (provided to readers).

Chapter 5 provides a formal introduction to exploratory data analysis (EDA) using the ggplot2 and tmap packages for visualising data and spatial data, respectively. EDA is a crucial component of any data analytics, allowing the univariate, pairwise and multivariate properties and structure of data to be explored and characterised. It builds on the visualisations and maps included in previous chapters, drawing them together to provide a coherent narrative around EDA and data visualisation.

Chapter 6 introduces some core ideas to support data analysis, with a focus on the development of models and testing their assumptions. It uses classical and Bayesian inference and suggests how these can be used to calibrate models, test hypotheses or provide predictions. This emphasis on modelling as a method of exploration aims to provide guidance for model construction in relation to data.

Chapter 7 builds on the ideas introduced in Chapter 6 and provides a formal treatment of data science, and the mechanics of machine learning for both inference and prediction, illustrating some classic machine learning approaches. It has a focus on the caret package and applies a number of caret implementations of machine learning algorithms to the same data to illustrate prediction, inference and classification.

Chapter 8 introduces some advanced alternative techniques for generating visual summaries of spatial data and the results of data analyses. These include cartograms, hexbins and tiling (pooling) which are applied to a local example before being applied to an analysis of the spatial trends in antidepressant prescribing ('the geography of misery') to illustrate how advanced visualisation approaches can be applied in practice. The assumptions behind each approach (cartograms, binning, aggregating over areas, etc.) are discussed.

Chapter 9 provides an epilogue. It reflects on the topics covered by the book as well as those that were touched on and require further consideration. It restates the book's underling principles for data analytics and the importance of taking a critical approach to data analysis and not just 'letting the data' speak as advocated by some. It concludes by describing a wish list of future development in data science that we feel would be beneficial and would fill the 'holes' in current tools, capabilities and concepts of spatial data science.

1.2.4 Learning and arcs

This book is aimed at both second- and third-year bachelor's/undergraduate students and postgraduate research students. The chapters sequentially build both in terms of content and in the complexity of the analyses they develop. By working through the illustrative code examples you will develop the skills to create your own routines, functions and programs. All of the chapters have a mix of introductory materials and advanced (and in places very detailed) analyses.

It is possible to create a number of different single-term courses or modules using this book for the computer lab/practical components. Each of the chapters is relatively self-contained in terms of concepts around which a lecture could be constructed (joining data, machine learning, databases, etc.) and the relevant computer lab/practical drawn from the book. Remember that the code for each chapter is on the book's website at https://study.sagepub.com/comber. The following single-term courses/modules could be constructed:

- 'An Introduction to Geographical Data Science' from Chapters 1 (Sections 1.3 and 1.4), 2, 3 and 5;
- 'Geographical Data Science in R' from Chapters 2, 3, 5 and 8;
- 'Advanced Techniques in Data Analytics in R' from Chapters 3, 4, 6 and 7;
- 'Advanced Geographical Data Science in R' from Chapters 3, 4, 7 and 8.

Each chapter should take around 3 hours of lab time for the 'average' student.

Elements of this book could also be used in combination with our other publication in this area (Brunsdon and Comber, 2018) which gives a deeper treatment of the various R data structures and formats (Chapter 2) and spatial data operations (Chapter 3). These provide a more comprehensive introduction to R for the novice user and could be incorporated into the 'Introduction to Geographical Data Science' and 'Geographical Data Science in R' outlines above. Chapter 4 in Brunsdon and Comber (2018) covers the use and creation of functions and structures commonly used in functions, and Chapter 5 in Brunsdon and Comber (2018) describes how to use R as a GIS. Chapters 6–9 in Brunsdon and Comber (2018) describe different types of spatial analysis comprehensively and could be used to augment the outlines for 'Advanced Techniques in Data Analytics in R' and 'Advanced Geographical Data Science in R' above.

The formal learning objectives of this book are:

- to support critical and technical understandings of spatial data analytics;
- to develop skills in handling and manipulating data and spatial data;
- to describe a suite of techniques for intelligently exploring the properties and structure (relationships) in data and spatial data;
- to provide an understanding of the relative advantages of different ways for analysing and visualising data to support the development of inferential theories;
- to develop skills in the R programming language.

Each chapter introduces a topic and has code snippets to run, with R scripts provided on the book's website. Earlier chapters provide the foundations for later ones. The dependencies and prerequisites for each chapter are fairly linear for the novice user: work through the chapters in sequence. The more advanced user will be able to go straight to whole chapters.

There is a strong emphasis placed on *learning by doing*, which means you are encouraged to unpick the code that you are given, adapt it and play with it. You are not expected to remember or recall each function you are introduced to. Instead you are encouraged to try to understand *what* operations you undertook, rather than *how* you undertook them. The learning curve is steep in places, and as you work though the book, the expectation is that you run all the code that you come across. We cannot emphasise enough that the best way to learn R and to become proficient in R coding is to write and enter it. Some of the code might look a bit intimidating when first viewed, especially in later chapters, but the only really effective way to understand it is to unpick each line.

The data required by the different chapters can be downloaded from https://study.sagepub.com/comber. The chapters requiring data contain links to the data to download them directly and instructions for loading the data in the R session.

1.3 GETTING STARTED IN R

1.3.1 Installing R and RStudio

We assume that most readers will be using the RStudio interface to R. The first thing you will need to do is install R. After this you can install RStudio. You should download the latest versions of R and RStudio in order to run the code provided in this book. At the time of writing, the latest version of R is version 3.6.3 and you should ensure you have at least this version. The simplest way to get R installed on your computer is to go the download pages on the R website – a quick search for 'download R' should take you there, but if not you could try:

- http://cran.r-project.org/bin/windows/base/ for Windows

- http://cran.r-project.org/bin/macosx/ for Mac

- http://cran.r-project.org/bin/linux/ for Linux

The Windows and Mac versions come with installer packages and are easy to install, while the Linux binaries require the use of a command terminal.

RStudio can be downloaded from https://www.rstudio.com/products/rstudio/download/ and the free distribution of RStudio Desktop is more than sufficient for this book (the version of RStudio is less important as this is essentially provides a wrapper for R). RStudio allows you to organise your work into projects, to use RMarkdown to create documents and webpages, to link to your GitHub site and much more. It can be customised for your preferred arrangement of the different panes.

1.3.2 The RStudio interface

We expect that most readers of this book and most users of R will be using the RStudio interface to R, although users can of course still use command-line R. RStudio provides a development environment with menus and window panes for different activities. RStudio has four panes, three of which show when you first open it: the **Console** on the left where code is entered (with a tab for Jobs); the working **Environment** with tabs for History and Connections; and **Files** with tabs for Plots, Packages, Help and Viewer. A fourth pane is opened when a script file of some kind is opened. Users can change the pane layout and set up their personal preferences for their RStudio interface.

Although RStudio provides a user-friendly interface, it is similar to straight R and has only a few pull-down menus. You will type or enter code in what is termed a *command-line interface* in the console window. Like all command-line interfaces, the learning curve is steep but the interaction with the software is more detailed, which allows greater flexibility and precision in the specification of commands.

Beyond this there are further choices to be made. Command lines can be entered in two forms: directly into the R console window or as a series of commands into a script window. We strongly advise that all code should be written in script (a .R file) and then run from the script (see below).

1.3.3 Working in R

You should start a new R session for each chapter. It is good practice to write your code in scripts and RStudio includes its own editor (similar to Notepad in Windows or TextEdit on a Mac). Scripts are useful if you wish to automate data analysis, and have the advantage of keeping a saved record of the relevant R programming language commands that you use in a given piece of analysis. These can be re-executed, referred to or modified at a later date. For this reason, you should get into the habit of constructing scripts for all your analysis. Since being able to edit functions is extremely useful, both the MS Windows and Mac OSX versions of R have built-in text editors. In RStudio you should go to **File > New File**. In R, to start the Windows editor with a blank document, go to **File > New Script** and to open an existing script **File > Open Script**. To start the Mac editor, use the menu options **File > New Document** to open a new document and **File > Open Document** to open an existing file.

Code may be written directly into a script or document as described below. If you have the PDF/ebook version of this book then you will be able to copy and paste the code into the script. However, copying and pasting from electronic sources can be affected by formatting issues. For this reason, access to the code in each chapter has been provided on the book's website (https://study.sagepub.com/comber). Snippets of code in your script can be highlighted (or the cursor placed against them) and then run, either by clicking on the Run icon at the top left of the script pane, or by pressing Ctrl-Enter (PC) or Cmd-Enter (Mac). Once code is written into these files, they can be saved for future use. RStudio also has a number of other keyboard shortcuts for running code, autofilling when you are typing, assignment, etc. Further tips are given at http://r4ds.had.co.nz/workflow-basics.html.

Scripts can be saved by selecting **File > Save As** which will prompt you to enter a name for the R script you have just created. Choose a name (e.g. test.R) and select **Save**. It is good practice to use the file extension .R.

The code snippets included in each chapter describe commands for data manipulation and analysis, to exemplify specific functionality. It is expected that you will run the R code yourself in each chapter. This can be typed directly into the R console but, to restate, we strongly advise that you create R scripts for your code. By creating scripts and running the code you will get used to using running code in the R console from the script. Comments in your script will help your understanding of the code's functionality (see below).

We advise that you create a working directory in Explorer (PC) or Finder (Mac) for each bit of work (project, chapter) that you do. It is also good practice to set the working directory at the beginning of your R session. This can be done via the menu in RStudio: **Session > Set Working Directory > Choose Directory**. Then you can create a new R script. Or, if you already have a script in that directory, choose **Session > Set Working Directory > To Source File Location**. In Windows R this is **File > Change dir...** and in Mac R this is **Misc > Set Working Directory**.

This points the R session to the folder you choose and will ensure that any files you wish to read, write or save are placed in this directory.

1.3.4 Principles

If you have never coded or used R before then working with scripts and command-line interfaces can seem a bit daunting. There is nothing we can say that will stop this, except that it will go away the more you do it! It is a bit like learning to play a musical instrument: it can feel like progress is slow, that the notation is in a foreign language (it is!) but there does come a point when you become familiar enough to able to run and write you own code. Like anything, the more you do it the better you become at it.

It is important to establish some general principles that we advocate in coding:

1. The code for all the analysis, maps, plots, tables, web maps, etc., is provided for you. The book's website has files containing the code used in each chapter (see https://study.sagepub.com/comber). The aim of this book is not to improve your typing, so you should *copy and paste code* to your script where possible before running it.

2. Learn to *interpret errors*. There are only a few things that can go wrong in any coding, and you will learn to interpret the errors if you read them. The most common cause of code not running and throwing up some kind of error message is because of typos. R is case sensitive (t is not the same as T), brackets of all kinds should be matched (i.e. open *and* closed) and different types of brackets do different things. Another common error is that the R objects that you pass to a function are not present in R's working environment. The internet (see below) can help, especially in unpicking errors, if you copy and paste the error message into an internet search engine.

3. You are expected to create your own R scripts to contain the code that you can then run in your session. It is good practice to write *comments* in your code. Comments are prefixed by #. Enter the two lines below into your R script and run it. Note that in the script, anything that follows a # on the command line is taken as comment and ignored by R. Including comments will help you especially when you return to a code snippet that you created a long time ago.

```
# this is a comment: assign a sequence of values to x
x = 5:12
```

4. In order to aid your *understanding* of what is being done by each code snippet, you should examine the *help* for individual functions. All functions have a help file that can be accessed using `help(<function_name>)` or `?<funtion_name>`, and in most cases an example of the function being applied is given. You should be able to simply highlight the example code and run it. Recall that code should can be highlighted (or the cursor placed against it) and then run, either by clicking on the Run icon at the top left of the script pane, or by pressing Ctrl-Enter (PC) or Cmd-Enter (Mac).

5. Having a *problem to solve* is by far the best way to learn coding. If you have a problem that you are trying to address, then you have a vested interest in the code (not just what we, the authors, thought was interesting or useful). You should break the problem down into discrete sub-tasks and then try to undertake each of these separately, building towards your answer. So as you are working through the book, think about how the examples relate to your analysis or study.

6. Use the *internet*. No source of information about R (or any other programming environment) is comprehensive. It does not matter whether you use a search engine to find R keywords (e.g. '"conditional" select rows "dplyr"') or specific webpages like https://stackoverflow.com/ (e.g. https://stackoverflow.com/search?q=%5Br%5D+dplyr). There is plenty of information out there, and the chances are that someone has solved your problem and published the answer; the tricky bit is to find the correct terminology to describe it!

1.4 ASSIGNMENT, OPERATIONS AND OBJECT TYPES IN R

This section introduces some basic concepts for working in R. It is about 20 pages or so in length and so can only provide a cursory introduction. It introduces some of the essentials of working in R, with a focus on basic data structures (vectors, matrices, data frames, etc.) and data types (integer, double, character, etc.). These are continually extended in subsequent chapters, with, for example, spatial data being introduced in Chapter 2. If you feel that you need further introductory materials after you have finished working through these then we recommend the following online *get started in R* guides:

- The Owen Guide (https://cran.r-project.org/doc/contrib/Owen-TheRGuide.pdf)

- 'An Introduction to R' (https://cran.r-project.org/doc/contrib/Lam-IntroductionToR_LHL.pdf)

- 'R for Beginners' (https://cran.r-project.org/doc/contrib/Paradis-rdebuts_en.pdf).

And of course there is our own offering, Brunsdon and Comber (2018), which provides a comprehensive introduction to data and spatial formats R in Chapters 2 and 3 (see https://uk.sagepub.com/en-gb/eur/an-introduction-to-r-for-spatial-analysis-and-mapping/book258267).

If you have never coded before or used R, then you should work your way carefully and slowly through this part of the chapter, using the principles listed in Section 1.3.4. Key among these are to always write code into an R script and to get used to examining the help for functions as they are introduced to you as a matter of routine.

1.4.1 Your first R script

This section describes the basic ways of assigning values to different *types* of R object (e.g. character, logical and numeric) and different *classes* of R object (e.g. vector, matrix, list, data.frame, factor). In describing these, different techniques for manipulating and extracting elements for different R objects are illustrated, along with some useful base functions.

R and RStudio must be installed on your system! We assume that you will be using RStudio.

You should create a working directory for this chapter. Open RStudio, open a new R script **File > New File > R Script** and save it to your working directory (**File > Save As**). You should notice that fourth pane now opens with a blank R script. Save this in your working directory, and give it an appropriate name (e.g. chap1.R).

Recall that you can create a working directory in two ways via the menu in RStudio. First, if you have not created a script, by **Session > Set Working Directory > Choose Directory**; then you can create a new R script. Second, if you already have a saved script in a directory, by **Session > Set Working Directory > To Source File Location**. This points the R session to the folder you choose and will ensure that any files you wish to read, write or save are placed in this directory.

In the command-line prompt in the console window, the > is an invitation to start typing in your commands. For example, type 2+9 into the console and press the Enter key:

```
2+9
## [1] 11
```

Here the result is 11. The [1] that precedes it formally indicates *first requested element will follow*. In this case there is just one element. The > indicates that R is ready for another command and all code outputs that are reproduced in this book are prefixed by '##' to help the reader differentiate these from code inputs.

So for the first bit of R code type the code below into your new R script:

```
y = c(4.3,7.1,6.3,5.2,3.2,2.1)
```

Notice the use of the c in the above code. This is used to *combine* or *concatenate* individual objects of the same type (numeric, character or logical – see below).

To run the code (in the console pane), a quick way is to highlight the code (with the mouse or using the keyboard controls) and press Ctrl-R or Cmd-Enter on a Mac. There is also a Run button at the top right of your script panel in RStudio: place your mouse cursor on the line and click on Run to run the code.

What you have done with the code snippet is *assign* the values to y, an R object. Assignment is the basic process of passing values to R objects, which are held in R's memory – if you look at the Environment pane in RStudio you will see that it now has an object called y. Assignment will generally be done using an = (equals sign), but you should note that it can also be done with a <- (less than, dash).

```
y <- c(4.3,7.1,6.3,5.2,3.2,2.1)
```

For the time being you can assume that *assignment* with = and <- are the same (although they are not, as will be illustrated with piping syntax in later chapters).

Now you can undertake *operations* on the objects held in the R working environment. You should write the following code into your R script and run each line individually or as a block by highlighting multiple lines and then running the code as before. Recall that highlighted code can be run by clicking on the Run icon at the top left of the script pane, or by pressing Ctrl-Enter (PC) or Cmd-Enter (Mac).

```
y*2
max(y)
```

There are two kinds of things here. The first is a mathematical operation (y*2). Operations do something to an R object directly, using mathematical notation like * for multiply. The second is the application of the function max to find the maximum value in an R object.

Recall that your understanding of what is being done by code snippets will grow if you explore the code, play around with it and examine the help for individual functions. All functions have a help file that can be accessed using help(<function_name>). Enter the code below into your script and run it to examine the help for max:

```
help(max)
```

Another way of doing this is to use ?

```
## alternative way to access help
?max
```

You should type the functions below into your R script and examine the help for these:

```
sum(y)
mean(y)
```

A key thing to note here is that *functions* are always followed by round brackets or *parentheses* (). These are different from square brackets (or *brackets*) [] and curly brackets (or braces) {}, as will be illustrated later. And functions (nearly!) always return something to the console and have the form:

```
result <- function_name(<input>)
```

Recall that it is good practice to write *comments* in your code. Comments are pre-fixed by # and are ignored by R when entered into the console. Type and enter the below into your R script and run it.

```
# assign 100 random, normally distributed values to x using rnorm
x <- rnorm(100)
```

You should remember to write your code and add comments as these will help you understand your code and what you did when you come back to it at a later date.

As hinted at above with the first code snippet adding 2 to 9, R can be used a bit like a calculator. It evaluates and prints out the result of any expression that is entered at the command line in the console. Recall that anything after a # prefix on a line is not evaluated. Type the code snippets below into your R script and run them:

```
2/9
sqrt(1000) # square root
2*3*4*5
pi # pi
2*pi*6378 # earth circumference
sin(c(1, 3, 6)) #sine of angles in radians
```

Key Points

- Code can (should) be run from a script.

- You should include comments in your scripts.

- Mathematical operations can be applied to R objects.

- Functions have round brackets or parentheses in the form `function_name(<input>)`.

- Functions take inputs, do something to them and return the result.

- Each function has a help page that can be accessed using `help(function_name)` or `?function_name`.

1.4.2 Basic data types in R

The first bit of coding above was to get you used to the R environment. You should have a script with a few code snippets. Now we will step back and examine in a bit more detail some of the structures and operations in R.

The preceding sections created two R objects, x and y – you should see them in the Environment pane in RStudio or by entering ls() at the console (this *lists* the objects in the R environment). There are a number of fundamental data types in R that provide the building blocks for data analysis. The sections below explore some of these data types and illustrate further operations on them.

1.4.2.1 Vectors

A vector is a group of values of the same type. The individual values are combined using c (); you have created one already. Examples of vectors are:

```
c(2,3,5,2,7,1)
3:10 # the sequence numbers 3, 4, ..., 10
c(TRUE,FALSE,FALSE,FALSE,TRUE,TRUE,FALSE)
c("London","Leeds","New York","Montevideo", NA)
```

Vectors may have different modes such as logical, numeric or character. The first two vectors above are numeric, the third is logical (i.e. a vector with elements of TRUE or FALSE values), and the fourth is a character or string vector (i.e. a vector with elements of mode character). The missing-value symbol, which is NA, can be included as an element of a vector.

Recall that the c in c(2, 3, 5, 2, 7, 1) above tells R to *combine* or *concatenate* the values in the parentheses. It tells R to *join these elements together into a vector.* Existing vectors may be included among the elements that are concatenated. In the following, vectors x and y are defined (overwriting the x and y that were defined earlier) and then concatenated to form a vector z:

```
x <- c(2,3,5,2,7,1)
x

## [1] 2 3 5 2 7 1

y <- c(10,15,12)
y

## [1] 10 15 12

z <- c(x, y)
z

## [1] 2 3 5 2 7 1 10 15 12
```

Vectors can be subsetted – that is, elements from them can be extracted. There are two common ways to extract subsets of vectors. Note in both cases the use of the square brackets [ ].

1. Specify the positions of the elements that are to be extracted, for example:

```
x[c(2,4)]  # Extract elements 2 and 4
## [1] 3 2
```

Note that negative numbers can be used to *omit* specific vector elements:

```
x[-c(2,3)]  # Omit elements 2 and 3
## [1] 2 2 7 1
```

2. Specify a vector of *logical values*. The elements that are extracted are those for which the logical value is TRUE. Suppose we want to extract values of x that are greater than 4:

```
x[x > 4]
## [1] 5 7
```

Examine the logical selection:

```
x > 4
## [1] FALSE FALSE   TRUE FALSE   TRUE FALSE
```

Further details on logical operations for extracting subsets are given later in this chapter.

1.4.2.2 Matrices and data frames

Matrices and data frames are like data tables with a row and column structure. The fundamental difference between a matrix and a data.frame is that matrices can only contain a single data type (numeric, logical, character, etc.) whereas a data frame can have different types of data in each column. All elements of any column must have the same type (e.g. all numeric).

Matrices are easy to define, but notice how the sequence 1 to 10 below is ordered differently with the byrow parameter:

```
matrix(1:10, ncol = 2)

##        [,1] [,2]
## [1,]    1    6
## [2,]    2    7
## [3,]    3    8
## [4,]    4    9
## [5,]    5   10
```

```
matrix(1:10, ncol = 2, byrow = T)
##         [,1] [,2]
## [1,]     1    2
## [2,]     3    4
## [3,]     5    6
## [4,]     7    8
## [5,]     9   10

matrix(letters[1:10], ncol = 2)
##         [,1] [,2]
## [1,]    "a"  "f"
## [2,]    "b"  "g"
## [3,]    "c"  "h"
## [4,]    "d"  "i"
## [5,]    "e"  "j"
```

Many R packages come with datasets. The iris dataset is an internal R dataset and is loaded to your R session with the code below:

```
data(iris)
```

This is a data.frame:

```
class(iris)
## [1] "data.frame"
```

The code below uses the head() function to print out the first six rows and the dim() function to tell us the dimensions of iris, in this case 150 rows and 5 columns:

```
head(iris)
##    Sepal.Length Sepal.Width Petal.Length Petal.Width Species
## 1           5.1         3.5          1.4         0.2  setosa
## 2           4.9         3.0          1.4         0.2  setosa
## 3           4.7         3.2          1.3         0.2  setosa
## 4           4.6         3.1          1.5         0.2  setosa
## 5           5.0         3.6          1.4         0.2  setosa
## 6           5.4         3.9          1.7         0.4  setosa

dim(iris)
## [1] 150    5
```

The str function can be used to indicate the formats of the attributes (columns, fields) in iris:

```
str(iris)
## 'data.frame':    150 obs. of   5 variables:
##  $ Sepal.Length: num  5.1 4.9 4.7 4.6 5 5.4 4.6 5 4.4 4.9 ...
##  $ Sepal.Width : num  3.5 3 3.2 3.1 3.6 3.9 3.4 3.4 2.9 3.1 ...
##  $ Petal.Length: num  1.4 1.4 1.3 1.5 1.4 1.7 1.4 1.5 1.4 1.5 ...
##  $ Petal.Width : num  0.2 0.2 0.2 0.2 0.2 0.4 0.3 0.2 0.2 0.1 ...
##  $ Species     : Factor w/ 3 levels "setosa","versicolor",..: 1 1
   1 1 1 1 1 1 ...
```

Here we can see that four of the attributes are numeric and the other is a factor –
a kind of ordered categorical variable.

The summary function is also very useful and shows different summaries of the
individual attributes in iris:

```
summary(iris)
##   Sepal.Length    Sepal.Width     Petal.Length    Petal.Width
##  Min.   :4.300   Min.   :2.000   Min.   :1.000   Min.   :0.100
##  1st Qu.:5.100   1st Qu.:2.800   1st Qu.:1.600   1st Qu.:0.300
##  Median :5.800   Median :3.000   Median :4.350   Median :1.300
##  Mean   :5.843   Mean   :3.057   Mean   :3.758   Mean   :1.199
##  3rd Qu.:6.400   3rd Qu.:3.300   3rd Qu.:5.100   3rd Qu.:1.800
##  Max.   :7.900   Max.   :4.400   Max.   :6.900   Max.   :2.500
##          Species
##  setosa    :50
##  versicolor:50
##  virginica :50
##
##
##
```

The main R graphics function is plot(), and when applied to a data frame or a
matrix it shows how the different attribute values (in fields or columns) correlate
to each other. A helpful graphical summary for the iris data frame is the scatter
plot matrix, shown in Figure 1.1 for the first four fields in iris. Note how the
inclusion of upper.panel=panel.smooth causes the lowess curves to be added to
Figure 1.1.

```
names(iris)
plot(iris[,1:4], pch = 1, cex = 0.7, col = "grey30",
     upper.panel=panel.smooth)
```

Notice also how the code above used plot(iris[,1:4]). This tells R to only
plot fields (columns) 1 to 4 in the iris data table (the numeric fields). Section 1.4.3

describes how to subset and extract elements from R objects in greater detail. The data types for individual fields can also be investigated using the `sapply` function:

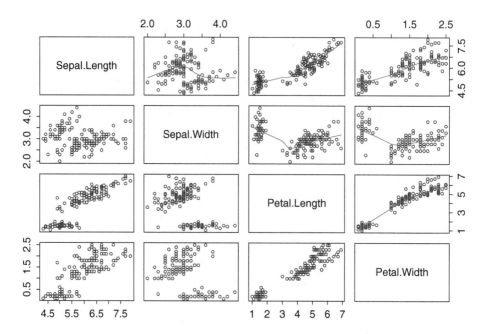

Figure 1.1 A plot of the numeric variables in the iris data

```
sapply(iris, class)
```

A key property in a data frame is that columns can be vectors of any mode. It is effectively a list (group) of column vectors, all of equal length. A matrix, on the other hand, requires all data elements to be of the same type. This is illustrated in the code below which shows the first 10 records from `iris` and what happens to them if they are coerced to `matrix` format with and without the fifth column (a factor variable):

```
iris[1:10,]
as.matrix(iris[1:10,])
as.matrix(iris[1:10,-5])
```

You should note that if the matrix contains a mix of data types, then it converts everything to character.

1.4.2.3 Factors

Factors provide a compact way to store character strings of categorical or classi-fied data. They are stored internally as a numeric vector with values 1, 2, 3, ..., *k*, where *k* is the number of levels, with an attributes table providing the *level* for each integer value.

The Species attribute of iris is a factor:

```
iris$Species
```

The function levels() allows factor levels to be inspected:

```
levels(iris$Species)
```

The attributes() shows the detail of the individual factors:

```
attributes(iris$Species)
## $levels
## [1] "setosa"     "versicolor" "virginica"
##
## $class
## [1] "factor"
```

The main advantages of factors are that values not listed in the levels cannot be assigned to the variable (try entering the code below), and that they allow easy grouping by other R functions as demonstrated in later chapters.

```
iris$Species[10] = "bananas"
```

There are many more data types in R. Chapter 2 in Brunsdon and Comber (2018) provides a good overview with worked examples and exercises.

Key Points

- A *vector* is a group of values of the same type.
- Vectors are created by combining individual values using c().
- R objects can be overwritten: the objects x and y were reassigned values in the examples above.
- Matrices and data frames are like data tables with a row and column structure.
- A matrix contains only a single data type (e.g. numeric, logical, character).

- A data frame can have different types of data in each column.

- R has many internal datasets such as the iris dataset.

- Factors are categorical or classified data elements and can be inspected with the levels() function.

1.4.3 Basic data selection operations

In the plot call above there are number of things to note (as well as the figure that was generated). In particular note the use of the vector 1:4 to index the columns of iris:

```
plot(iris[,1:4], pch = 1, cex = 1.5)
```

Notice how 1:4 extracted all the columns between 1 and 4. Effectively this code says: plot columns 1, 2, 3 and 4 of the iris dataset, using plot character (pch) 1 and a character expansion factor (cex) of 1.5. This is very useful and hints at how elements of vectors, flat data tables such as matrices and data frames can be extracted and/or ordered by using square brackets.

Starting with vectors, the code below extracts different elements from z that you defined earlier:

```
# z
z
# 1st element
z[1]
# 5th element
z[5]
# elements 1 to 6
z[1:6]
# elements 1, 3, 6 and 2 ... in that order
z[c(1,3,6,2)]
```

So for vectors, a vector of numbers in a square bracket can be used to extract elements.

For two-dimensional data tables (like data frames, matrices, tibbles, spatial data tables, etc., that you will encounter as you work through this book) a two-dimensional notation within the square brackets is required. This is in the form of [<rows>, <columns>]. Run the code below:

```
names(iris)
names(iris)[c(1,4)]
names(iris)[c(4,2)]
plot(iris[,c(3,4,2,1)])
```

Notice a number of things:

- In the second line above, the first and fourth elements from the vector of column names were selected.

- In the third line the fourth and second elements were extracted.

- For the plot, the vector was passed to the second argument, after the comma, in the square brackets [,] to indicate which columns were to be plotted.

The referencing or *indexing* in this way is very important: individual rows and columns of two-dimensional data structures such as data frames, matrices and tibbles can be accessed by passing references to them in the square brackets:

```
# 1st row
iris[1,]
# 3rd column
iris[,3]
# a selection of rows
iris[c(3:7,10,11),]
```

Such indexing could of course have been assigned to an R object and used to do the subsetting:

```
x = c(1:3) # assign index to x
names(iris)[x] # check
plot(iris[,x], pch = 1, cex = 1.5)
```

Specific rows and columns can be selected using a pointer or an *index* which can be specified in three main ways:

- numerically – the code below returns the first 10 rows and 2 columns:

```
iris[1:10, c(1,3)]
```

- by name – the code below returns the first 10 rows and 2 named columns:

```
iris[1:10, c("Sepal.Length", "Petal.Length")]
```

- logically – the code below returns the first 10 rows and 2 logically selected columns:

```
iris[1:10, c(TRUE, FALSE, TRUE, FALSE, FALSE, FALSE)]
```

Thus there are multiple ways in which the ith rows or jth columns in a data table can be accessed. Also note that compound logical statements can be used to create an index as in the code below:

```
n <- iris$Sepal.Length > 6 & iris$Petal.Length > 5
iris[n,]
```

Arithmetic operations that may be used in the extraction of subsets of vectors are < <= > >= == !=. The first four compare magnitudes, == tests for equality, and != tests for inequality. We will cover logical operations in more detail in subsequent chapters.

In summary, there are two common ways to extract subsets of vectors and matrices:

1. Specify the elements that are to be extracted directly, for example:

```
z[c(2,4)] # Extract elements (rows) 2 and 4
## [1] 3 2
iris[, "Petal.Width"]
##    [1] 0.2 0.2 0.2 0.2 0.2 0.4 0.3 0.2 0.2 0.1 0.2 0.2 0.1 0.1 0.2
##   [16] 0.4 0.4 0.3 0.3 0.3 0.2 0.4 0.2 0.5 0.2 0.2 0.4 0.2 0.2 0.2
##   [31] 0.2 0.4 0.1 0.2 0.2 0.2 0.2 0.1 0.2 0.2 0.3 0.3 0.2 0.6 0.4
##   [46] 0.3 0.2 0.2 0.2 0.2 1.4 1.5 1.5 1.3 1.5 1.3 1.6 1.0 1.3 1.4
##   [61] 1.0 1.5 1.0 1.4 1.3 1.4 1.5 1.0 1.5 1.1 1.8 1.3 1.5 1.2 1.3
##   [76] 1.4 1.4 1.7 1.5 1.0 1.1 1.0 1.2 1.6 1.5 1.6 1.5 1.3 1.3 1.3
##   [91] 1.2 1.4 1.2 1.0 1.3 1.2 1.3 1.3 1.1 1.3 2.5 1.9 2.1 1.8 2.2
## [106] 2.1 1.7 1.8 1.8 2.5 2.0 1.9 2.1 2.0 2.4 2.3 1.8 2.2 2.3 1.5
## [121] 2.3 2.0 2.0 1.8 2.1 1.8 1.8 1.8 2.1 1.6 1.9 2.0 2.2 1.5 1.4
## [136] 2.3 2.4 1.8 1.8 2.1 2.4 2.3 1.9 2.3 2.5 2.3 1.9 2.0 2.3 1.8
```

You can also use negative numbers to omit elements:

```
z[-c(2,3)]
## [1]  2  2  7  1  10  15  12
```

2. Specify an expression that generates a vector of logical values. The elements that are extracted are those for which the logical value is TRUE. The code below extracts values of z that are greater than 10:

```
z[z > 10]
## [1] 15 12
```

The use of logical operators is given a more complete treatment in the next subsection.

Key Points

- Square brackets [] are used to indicate the locations of data elements in R objects.

- Referencing data locations in this way can be used to select data elements from R objects.

- For vectors, a vectors of numbers in square brackets extracts elements (e.g. x[c(1,2,4)]).

- For two-dimensional data tables (data frames, matrices, tibbles, etc.), a two-dimensional notation within the square brackets is used to extract elements, in the form of [<rows>, <columns>]. For example:

```
# 1st row
iris[1,]
# 2nd and 4th column
iris[, c(2,4)]
```

1.4.4 Logical operations in R

Two of the code snippets above used logical operations to extract elements from a vector. It is instructive to examine these in more detail. First run the code again to re-familiarise yourself with what is being done:

```
x[x > 4]
z[z > 10]
```

What these do in each case is print out (extract) the elements from x and z that are greater than 4 and 10, respectively. We can examine what is being done by the code in the square brackets and see that they return a series of logical statements. Considering x, we can see that it has six elements and two of these satisfy the logical condition of 'being greater than 4'. The code x > 4 also returns a vector with six elements but it is a logical one containing TRUE or FALSE values:

```
x
## [1] 2 3 5 2 7 1
x > 4
## [1] FALSE FALSE  TRUE FALSE  TRUE FALSE
```

Additionally, we can see that these are elements 3 and 5 of x. The code below combines x and the logical statement into a data frame to show these two vectors together:

```
data.frame(x = x, logic = x>4)
## x    logic
## 1 2 FALSE
## 2 3 FALSE
## 3 5  TRUE
## 4 2 FALSE
## 5 7  TRUE
## 6 1 FALSE
```

So effectively what the logical statement is doing is telling R to extract the third and fifth elements from x. It is the same as entering:

```
x(c(3,5))
```

And this is reinforced by using the function which with the logical statement. This returns the vector (element) positions for which the logic statement is TRUE:

```
which(x > 4)
## [1] 3 5
```

A further consideration is that logical statements can be converted to binary values of 0 or 1 very easily:

```
(x > 4) + 0
## [1] 0 0 1 0 1 0
(x > 4) * 1
## [1] 0 0 1 0 1 0
```

And logical statements can be inverted using the *not* or negation syntax (!) placed in front of the statement:

```
x > 4
## [1] FALSE FALSE  TRUE FALSE  TRUE FALSE
!(x > 4)
## [1] TRUE TRUE  FALSE TRUE  FALSE TRUE
```

So, returning to the original code snippets, what these do is use the logical statement applied to the whole vector, to return just the elements from that vector that match the condition. This could be done in long hand by creating an intermediate variable:

```
my.index = x > 4
x[my.index]
```

```
x[my.index]
```

Logical statements can also be used to extract or subset from two-dimensional data tables such as matrices and data frames. The code below prints out the rows from iris that have a Sepal.Length value of 7 or more:

```
iris[iris$Sepal.Length >= 7, ]
```

This is telling R to extract the rows for which the statement iris$Sepal.Length >= 7 is TRUE and does not specify any columns, so they are all returned. Have a look at how this works:

```
iris$Sepal.Length >= 7
```

So out of 150 records (rows) in the iris dataset, 13 have Sepal.Length attributes that are greater than or equal to 7. You could count the TRUE values and the rows in iris manually or run the code below:

```
# rows in iris
nrow(iris)
```

```
## [1] 150
```

```
# count of TRUE
sum(iris$Sepal.Length >= 7 )
```

```
## [1] 13
```

It is useful to understand the different arithmetic relations that can be used in these expressions, and for example to generate indexes. These include:

- greater than: >
- less than: <
- equals: == (note the double equals sign)
- not equal to: !=
- greater than or equal to: >=
- less than or equal to: <=

Different binary operators for generating logical statements can be found at help(Comparison).

It is also possible to generate *compound* logical statements composed of more than one condition. Different logical statements are linked by operators, the most commonly used being AND for multiple statements to be TRUE and OR for at least one statement to be TRUE. Some examples of these are illustrated in the code snippets below. Notice how the last of these is a single line of code but spread over two lines, with an indentation on the second line.

```
# Sepal.Length between 7 and 7.2 inclusive
iris[iris$Sepal.Length >= 7.0 & iris$Sepal.Length <= 7.2,]
# Sepal.Length greater than 7.0 and Petal.Length less than 6.0
iris[iris$Sepal.Length > 7.0 & iris$Petal.Length < 6.0,]
# Sepal.Length greater than 5.5 less than or equal to 5.7
iris[iris$Sepal.Length > 5.5 & iris$Sepal.Length <= 5.7,]
# Sepal.Length greater than 5.5 less than or equal to 5.7 AND
# Species is virginica OR setosa - notice the use of the brackets
iris[(iris$Sepal.Length > 5.5 & iris$Sepal.Length <= 5.7) &
        (iris$Species == "virginica" | iris$Species == "setosa"), ]
```

Similarly, the syntax for compound logical statements should be explored – you will principally be interested in AND (&) and OR (|):

```
?Syntax
```

Key Points

- Logical statements return a vector of TRUE and FALSE elements.

- These can be converted to binary [0, 1] format.

- Logical statements can be used to extract elements from one-dimensional and two-dimensional data either directly or by assigning the results of the statement to an R object that is later used to *index* the data.

- Compound logical statements can be constructed to specify a series of conditions that have to be met.

1.4.5 Functions in R

There are a number of useful functions in R as listed below. You should explore the help for these.

```
length(z)
mean(z)
median(z)
range(z)
unique(z)
sort(z)
order(z)
sum(z)
cumsum(z)
cumprod(z)
rev(z)
```

These operate over individual vectors. Functions can be applied to all columns of a data frame or matrix using the function sapply(). It takes as its arguments the data frame, and the function that is to be applied. The following applies the function is.factor() to all columns of the supplied data frame iris:

```
sapply(iris, is.factor)
## Sepal.Length  Sepal.Width Petal.Length Petal.Width   Species
##        FALSE         FALSE        FALSE       FALSE      TRUE
```

In this case, such TRUE and FALSE statements could be used as a method of indexing. The code below determines which of the columns of iris is a factor assigning the result to index and then uses the negation of that to subset the dataset passed to range.

```
index = sapply(iris, is.factor)
sapply(iris[,!index], range) # The first 4 columns are not factors

##      Sepal.Length Sepal.Width Petal.Length  Petal.Width
## [1,]          4.3         2.0          1.0          0.1
## [2,]          7.9         4.4          6.9          2.5
```

Note the use of ! to negate index – you could examine its effect!

```
index
!index
```

Two simple examples of R functions are described below. The first gives an approximate conversion from miles to kilometres:

```
miles.to.km <- function(miles)miles*8/5
# Distance from Maynooth to Nottingham is ~260 miles
miles.to.km(260)
```

The function will do the conversion for several distances all at once. To convert a vector of the three distances 100, 200 and 300 miles to distances in kilometres, specify:

```
miles.to.km(c(100,200,300))
```

Here is a function that makes it possible to plot the figures for any pair of variables in the data.

```
plot_iris <- function(x, y){
    x <- iris[,x]
    y <- iris[,y]
    plot(x, y)
}
```

Note that the function body is enclosed in curly brackets or braces ({ }).

```
plot_iris(1,2)
plot_iris("Sepal.Length", 3)
plot_iris("Sepal.Length", "Petal.Width")
```

A number of functions will be defined and used in later chapters throughout the book.

Key Points

- Functions are an integral part of using R. We do not have to re-invent the wheel ourselves – rather we can use the tools created by others.

- Functions are defined using braces ({}) and are called with parameters in parentheses in the form function_name(<input>) as described earlier.

- sapply applies a specified function to each column of a data table.

1.4.6 Packages

Finally, everything that you have done thus far in this chapter has worked with R's default tools and data that come with any R installation. This is R's *base functionality*. However, one of the key advantages of working in R is the ability to use packages created by others in the R community. These are libraries and, instead of books, contain functions and data. One of the joys of R is the community of users. Users share what they do and create in R in a number of ways. One of these is through packages. Packages are collections of related functions that have been created, tested and supported with help files. These are bundled into a package and shared with other R users via the CRAN repository. There are thousands of packages in R. The number of packages is continually growing. When packages are installed these can be called as libraries. The background to R, along with documentation and information about packages as well as the contributors, can be found at the R Project website (http://www.r-project.org).

Users need to install a package *only once* to mount it on their computer, and then it can be called in R scripts as required. The basic operations are as follows:

- Install the package before it is used for the first time (note that you may have to set a mirror site the first time you install a package). This is only done once.

- Load the package using the library() function to use the package tools. This is done for each R session.

In RStudio, packages can be installed in two ways. First, via the RStudio menu: **Tools > Install Packages…** This opens a dialogue box that allows you to enter the names of packages (remember they are case sensitive), either individually or in a comma-separated list. For example, to install the tidyverse package, you could enter tidyverse (note that as you type it starts to autofill), and it is good practice to check the Install dependencies tickbox.

Second, packages can be installed in the console using the install.packages function, which has the following syntax, again with the dependencies required by the package:

```
install.packages("package_name", dependencies = TRUE)
```

For example, to install the tidyverse package enter:

```
install.packages("tidyverse", dependencies = TRUE)
```

Whichever way you do this, via the menu system or via the console, you may have to respond to the request from R/RStudio to set a mirror – a site from which to download the package – pick the nearest one! Again a mirror only needs to be set once as RStudio will remember your choice.

You should install the tidyverse package now using one of the methods above, making sure that this is done with the package dependencies.

Dependencies are needed in most package installations. This is because most packages build on the functions contained in other packages. If you have installed tidyverse you will see that it loads a number of other packages. This is because the tidyverse package uses functions as building blocks for its own functions, and provides a *wrapper* for these (see https://www.tidyverse.org). These dependencies are loaded if you check the Install dependencies tickbox in the system menu approach or by including the dependencies = TRUE argument in the code above. In both cases these tell R to install any other packages that are required by the package being installed.

Once installed, the packages do not need to installed again for subsequent use. They can simply be called using the library function as below, and this is usually done at the start of the R script:

```
library(tidyverse)
```

A key point here is that packages are occasionally completely rewritten, and this can impact on code functionality. For example, since we started writing this book the read and write functions for spatial data in the maptools package (readShapePoly, writePolyShape, etc.) have deprecated.

Run the code below to install maptools with its dependencies (noting that this can be abbreviated to dep = TRUE):

```
install.packages("maptools", dep = TRUE)
library(maptools)
```

Now examine the help for the readShapePoly function using the help function:

```
help(readShapePoly)
```

Here you can see that the help pages for deprecated functions contain a warning and suggest other functions that should be used instead. The code on the book's website (https://study.sagepub.com/comber) will always contain up-to-date code snippets for each chapter to overcome any problems caused by function deprecation.

Such changes are only a minor inconvenience and are part of the nature of a dynamic development environment provided by R in which to do research: such changes are inevitable as packages finesse, improve and standardise. Further descriptions of packages, their installation and their data structures are given in later chapters.

Key Points

- R/RStudio comes with a large number of default tools.

- These can be expanded by installing user-contributed packages of tools and data.

- Packages only have to installed once (via the menu or the console), and then can be called subsequently using the library(package_ name) syntax.

- Packages should be installed with their dependencies (these are other packages that the target package builds on).

1.5 SUMMARY

The aim of Section 1.4 was to introduce you to R if you have not used it before, to familiarise you with the R/RStudio environment and to provide a basic grounding in R coding. You should have a script with all your R code and comments (comments are really important) describing how to assign values to R objects and some basic operations on those objects' data. You should understand some of the different basic data strictures and how to handle and manipulate them, principally one-dimensional vectors and two-dimensional matrices and data frames. Logical operations were introduced. These allow data elements to be extracted or subsetted. Functions were introduced, and you should have installed the `tidyverse` package. Some additional resources were suggested for those needing to spend a bit more time becoming more familiar with R. A number of excellent online *get started in R* guides were listed and the book *An Introduction to R for Spatial Analysis and Mapping* (Brunsdon and Comber, 2018) was recommended if a deeper introduction to data formats in R was needed.

REFERENCES

Brunsdon, C. and Comber, A. (2020a) Big issues for big data. Preprint arXiv:2007.11281.

Brunsdon, C. and Comber, A. (2020b) Open in a practice: Supporting reproducibility and critical spatial data science. *Journal of Geographical Systems*, https://doi.org/10.1007/s10109-020-00334-2.

Brunsdon, C. and Comber, L. (2018) *An Introduction to R for Spatial Analysis and Mapping* (2nd edn). London: Sage.

Comber, A., Brunsdon, C., Charlton, M. and Harris, R. (2016) A moan, a discursion into the visualisation of very large spatial data and some rubrics for identifying big questions. In *International Conference on GIScience Short Paper Proceedings*. Vol. 1, Issue 1.

Kitchin, R. (2013) Big data and human geography: Opportunities, challenges and risks. *Dialogues in Human Geography*, 3(3), 262–267.

Kitchin, R. and Lauriault, T. (2014) Towards critical data studies: Charting and unpacking data assemblages and their work. The Programmable City Working Paper 2, Preprint. https://ssrn.com/abstract=2474112.

Kitchin, R. and McArdle, G. (2016) What makes big data, big data? Exploring the ontological characteristics of 26 datasets. *Big Data & Society*, 3(1).

Laney, D. (2001) 3D data management: Controlling data volume, velocity and variety. META Group Research Note, 6 February.

Marr, B. (2014) Big data: The 5 Vs everyone must know. *LinkedIn Pulse*, 6 March.

McNulty, E. (2014) Understanding big data: The seven V's. *Dataconomy*, 22 May.

Myers, J. L., Well, A. D. and Lorch, R. F. Jr. (2013) *Research Design and Statistical Analysis*. New York: Routledge.

O'Neil, C. and Schutt, R. (2014) *Doing Data Science: Straight Talk from the Front Line*. Sebastopol, CA: O'Reilly.

Openshaw, S. (1984a) Ecological fallacies and the analysis of areal census data. *Environment and Planning A*, 16(1), 17–31.

Openshaw, S. (1984b) *The Modifiable Areal Unit Problem*, Catmog 38. Norwich: Geo Abstracts.

Pebesma, E., Bivand, R., Racine, E., Sumner, M., Cook, I., Keitt, T. et al. (2019). Simple features for R. https://cran.r-project.org/web/packages/sf/vignettes/sf1.html.

Robinson, W. (1950) Ecological correlations and the behavior of individuals. *American Sociological Review*, 15, 351–357.

Tobler, W. R. (1970) A computer movie simulating urban growth in the Detroit region. *Economic Geography*, 46 (sup1), 234–240.

Wickham, H., Averick, M., Bryan, J., Chang, W., McGowan, L., François, R., Grolemund, G. et al. (2019) Welcome to the Tidyverse. *Journal of Open Source Software*, 4(43), 1686.

(2)

DATA AND SPATIAL
DATA IN R

2.1 OVERVIEW

This chapter introduces data and spatial data. It covers recent developments in R that have resulted in new formats for data (`tibble` replacing `data.frame`) and for spatial data (the `sf` format replacing the `sp` format). The chapter describes a number of issues related to data generally. including the structures used to store data and *tidy* data manipulations with the `dplyr` package, and introduces the `tmap` package for mapping and visualising spatial data properties, before they are given a more comprehensive treatment in later chapters (Chapters 3 and 5, respectively).

The chapter introduces some data-table-friendly functions and some of the critical issues associated with *spatial* data science, such as how to move between different spatial reference systems so that data can be spatially linked. Why spatial data? Nearly all data are *spatial*: they are collected some*where*, at some *place*. In fact this argument can probably be extended to the spatio-temporal domain: all data are spatio-temporal – they are collected some*where* and at some *time* (Comber and Wulder, 2019).

This chapter covers the following topics:

- Data and spatial data
- The `tidyverse` package and tidy data
- The `sf` package and spatial data
- The `dplyr` package for manipulating data.

It is expected that the reader has a basic understanding of data formats in R, and has at least worked their way through the introductory materials in Chapter 1. They should be familiar with assigning values to different types of R object (e.g. `character`, `logical` and `numeric`) and different classes of R object (e.g. `vector`, `matrix`, `list`, `data.frame` and `factor`). If you have worked through our other offering (Brunsdon and Comber, 2018) then some of this chapter will be revision.

The following packages will be required for this chapter:

```
library(tidyverse)
library(sf)
library(tmap)
library(sp)
library(datasets)
```

If you have not already installed these packages, this can be done using the `install.packages()` function as described in Chapter 1. For example:

```
install.packages("tidyverse", dep = T)
```

Note that if you encounter problems installing packages then you should try to install them in R (outside of RStudio) and then call the now installed libraries in RStudio.

Remember that you should write your code into a script, and set the working directory for your R/RStudio session. To open a new R script select **File > New File > R Script** and save it to your working directory (**File > Save As**) with an appropriate name (e.g. chap2.R). Having saved your script, you should set the working directory. This can be done via **Session > Set Working Directory > Choose Directory**. Or, if you already have a saved script in a directory, select **Session > Set Working Directory > To Source File Location**. This should be done each time you start working on a new chapter.

2.2 DATA AND SPATIAL DATA

2.2.1 Long vs. wide data

This section starts by examining data formats, comparing non-spatial or *aspatial* data with spatial data. However, it is instructive to consider what is meant by *data* in this context. R has many different ways of storing and holding data, from individual data elements to lists of different data types and classes.

The commonest form of data is a flat data table, similar in form to a spreadsheet. In this *wide* format, each of the rows (or *records*) relates to some kind of observation or real-world feature (e.g. a person, a transaction, a date, a shop) and the columns (or *fields*) represent some attribute or property associated with that observation. Each cell in wide data tables contains a value of the property of a single record.

Much *spatial* data has a similar structure. Under the object or *vector* form of spatial data, records still refer to single objects, but these relate to some real-world *geographical* feature (e.g. a place, a route, a region) and the fields describe variables, measurements or attributes associated with that feature (e.g. population, length, area).

Data can also be *long* or flat. In this format, the observation or feature is retained but the multiple variable fields are collapsed typically into three columns: one containing observation IDs or references, another describing the variable name, domain or type, and the third containing the value for that observation in that domain. Long data will have different levels of *longness* depending on the number of variables.

Depending on the analysis you are undertaking, you may require wide format data or long format data. Data are usually in wide format, but for some data science activities long format data are more commonly required than wide format data, particularly the ggplot2 package for data visualisations. Thus it is important to be able to move between wide and long data formats.

The tidyr package provides functions for doing this. The pivot_longer() function transforms from wide to long format, and the pivot_wider() function transforms from long to wide format.

The code below generates wide and long format data tables from a subset of the first five records and of the first three fields of the mtcars dataset, one of the datasets automatically included with R. The results are shown in Tables 2.1 and 2.2.

```
# load the data
data(mtcars)
# create an identifier variable, ID, from the data frame row names
mtcars$ID = rownames(mtcars)
# sequentially number the data frame rows
rownames(mtcars) = 1:nrow(mtcars)
# extract a subset of the data
mtcars_subset = mtcars[1:5,c(12, 1:3)]
# pivot to long format
mtcars_long = pivot_longer(mtcars_subset, -ID)
# pivot to wide format
mtcars_wide = pivot_wider(mtcars_long)
mtcars_long
mtcars_wide
```

Table 2.1 The wide data table format

ID	mpg	cyl	disp
Mazda RX4	21.0	6	160
Mazda RX4 Wag	21.0	6	160
Datsun 710	22.8	4	108
Hornet 4 Drive	21.4	6	258
Hornet Sportabout	18.7	8	360

Table 2.2   The long data table format

ID	name	value
Mazda RX4	mpg	21.0
Mazda RX4	cyl	6.0
Mazda RX4	disp	160.0
Mazda RX4 Wag	mpg	21.0
Mazda RX4 Wag	cyl	6.0
Mazda RX4 Wag	disp	160.0
Datsun 710	mpg	22.8
Datsun 710	cyl	4.0
Datsun 710	disp	108.0
Hornet 4 Drive	mpg	21.4
Hornet 4 Drive	cyl	6.0
Hornet 4 Drive	disp	258.0
Hornet Sportabout	mpg	18.7
Hornet Sportabout	cyl	8.0
Hornet Sportabout	disp	360.0

Key Points

- Data tables are composed of rows or records, and columns or fields.
- Flat data tables have a row for each observation (a place, a person, etc.) and a column for each variable (field).
- Long data format contains a record for each unique observation–field combination.

2.2.2 Changes to data formats

The standard formats for tabular data and vector-based spatial data have been the data.frame and sp formats. However, R is a dynamic coding and research environment: things do not stay the same. New tools, packages and formats are constantly being created and updated to improve, extend and increase the functionality and consistency of operations in R. Occasionally a completely new paradigm is introduced, and this is the case with the recent launch of the tibble and sf data formats for data and spatial data, respectively.

The tibble format was developed to replace the data.frame. It is a reimagining of the data frame that overcomes the limitations of the data.frame format.

The emergence of the `tibble` as the default format for data tables has been part of a coherent and holistic vision for data science in R supported by the `tidyverse` package (https://www.tidyverse.org). This has been transformatory: `tidyverse` provides a collection of integrated R packages supporting data science that have redefined data manipulation tools, vocabularies and functions as well as data formats (Wickham, 2014). The `tibble` package is loaded with `dplyr` as part of `tidyverse`. Without the `tidyverse` package, this book could not have been written in the way that it has been. The `tidyverse` and `tibble` format have also underpinned a parallel development in *spatial* data. At the time of writing the `sp` spatial data format is in the process of being replaced by `sf` or the *simple feature* data format for vector spatial data. These developments are described in the next sections.

2.2.3 Data formats: `tibble` vs. `data.frame`

The `tibble` class in R is composed of a series of `vectors` of equal length, which together form two-dimensional data structures. Each vector records values for a particular variable, theme or attribute and has a name (or *header*) and is ordered such that the *n*th element in the vector is a value for the *n*th record (row) representing the *n*th feature. These characteristics also apply to the `data.frame` class, which at the time of writing is probably the most common data format in R.

The `tibble` is a reworking of the `data.frame` that retains its advantages (e.g. multiple data types) and eliminates less effective aspects (some of which are illustrated below). However, the `tibble` reflects a tidy (Wickham, 2014) data philosophy:

- It allows *multiple types* of variable or attribute to be stored in the same table (unlike, for example, the `matrix` format which can only hold one data *type*, such as `integer`, `logical` or `character`).

- It seeks to be *lazy* and does not do any work trying, for example, to link partially matched variable names (unlike the `data.frame` format – see the example below).

- It is *surly* and complains if calls to the data are not exactly specified, identifying problems earlier in the data analysis cycle and thereby forcing cleaner coding.

By contrast, the `data.frame` format is not *tidy*: it is not *lazy* or *surly*. This is illustrated in the code snippets below. These use the `mtcars` dataset loaded above (a `data.frame`) to create a new `data.frame` and then creates a `tibble` from this to highlight the differences between the two formats.

First, create a `data.frame` called `df` from `mtcars`:

```
# create 4 variables
type = as.character(mtcars$ID)
weight = mtcars$wt
horse_power = mtcars$hp
q_mile <- mtcars$qsec
# create a new data.frame
df <- data.frame(type, weight, horse_power, q_mile)
```

Then the structure of `df` can be examined using the `str()` function:

```
str(df)

## 'data.frame':  32 obs. of 4 variables:
## $ type       : Factor w/ 32 levels "AMC Javelin",..: 18 19 5 13
14 31 7 21 20 22 ...
## $ weight     : num 2.62 2.88 2.32 3.21 3.44 ...
## $ horse_power: num 110 110 93 110 175 105 245 62 95 123 ...
## $ q_mile     : num 16.5 17 18.6 19.4 17 ...
```

The `type` attribute has been converted from a `character` format to a `factor` format by the `data.frame()` function: by default it encodes character strings into factors. This can be seen in the `str` function above and by the `Levels` that are indicated in the code below:

```
unique(df$type)
```

To overcome this the `df` object can be refined using `stringsAsFactors = FALSE`:

```
df <- data.frame(type, weight, horse_power, q_mile,
                 stringsAsFactors = FALSE)
str(df)

## 'data.frame':  32 obs. of 4 variables:
## $ type       : chr "Mazda RX4" "Mazda RX4 Wag" "Datsun 710"
"Hornet 4 Drive"...
## $ weight     : num 2.62 2.88 2.32 3.21 3.44 ...
## $ horse_power: num 110 110 93 110 175 105 245 62 95 123 ...
## $ q_mile     : num 16.5 17 18.6 19.4 17 ...
```

Next, create a `tibble` called `tb` using the code below:

```
tb <- tibble(type, weight, horse_power, q_mile)
```

And examine its structure:

```
str(tb)
```

Here we can see that the character values in type have not been converted to a factor. However, probably the biggest criticism of data.frame is the partial matching behaviour. Enter the following code:

```
head(df$ty)
head(tb$ty)
```

Although there is no variable called ty, the partial matching in the data.frame means that the type variable is returned. This is a bit worrying!

A further problem is what gets returned when a data table is subsetted. A tibble always returns a tibble, whereas a data.frame may return a vector or a data.frame, depending on the dimensions of the result. For example, compare the outputs of the following code:

```
# a single column - the second one
head(df[,2])
head(tb[,2])
class(df[,2])
class(tb[,2])
# the first 2 columns
head(df[,1:2])
head(tb[,1:2])
class(df[,1:2])
class(tb[,1:2])
```

Some final considerations are that the print method for tibble returns the first 10 records by default, whereas for a data.frame all records are returned. Additionally, the tibble class includes a description of the class of each field (column) when it is printed. Examine the differences between these data table formats:

```
tb
df
```

It is possible to convert between a tibble and data.frame using the following functions:

```
data.frame(tb)
as_tibble(df)
```

You should examine the tibble vignettes and explore their creation, coercion, subsetting and so on:

```
vignette("tibble")
```

A quick note about vignettes. Vignettes are short descriptive documents that contain worked examples of a particular set of techniques. They allow the package authors to provide in-depth narratives in order to support R users in the application of the tools and concepts embedded in the package. Not all packages include vignettes, but they are increasingly used as a way of providing background explanations to, for example, new concepts such as *simple features* (sf package) and *tibbles* (tibble package within tidyverse).

In this book, the tibble format is used as the default for aspatial or non-spatial data tables (although some examples use the data.frame format) because it has been designed to support data manipulations with the dplyr package (Chapter 3), data visualisations using the ggplot2 package (Chapters 5 and 8) as well as data analysis (Chapters 4, 6 and 7) using spatial queries and data extracted from remote databases.

Finally, and for completeness, data tables, in tibble or data.frame formats, can be easily read into R and written out to a local file. The base installation of R comes with core functions for reading and writing .txt, .csv and other tabular formats to save and load data to and from local files.

The code below writes df to a CSV file in the current working directory:

```
write.csv(df, file = "df.csv", row.names = FALSE)
```

The CSV file can be read in with the stringsAsFactors parameter set to FALSE to avoid the factor issue with the data.frame format described above:

```
df2 = read.csv("df.csv", stringsAsFactors = F)
str(df2)
```

The write.table() function can be used to write TXT files with different field separations:

```
# write a tab-delimited text file
write.table(df, file = "df.txt", sep = "\t", row.names = F,
            qmethod = "double")
df = read.table(file = "df.txt", header = T, sep= "\t",
                stringsAsFactors = F)
head(df2)
str(df2)
```

Data tables in tibble format can be treated in a similar way but using the read_csv() and write_csv() functions:

```
tb2 = read_csv("df.csv")
write_csv(tb, "tb.csv")
```

```
# write a tab-delimited text file
write_delim(tb, path = "tb.txt", delim = "\t",
            quote_escape = "double")
```

The advantages and hints for effective reading of tidy data are described more fully in Section 3.4 of Chapter 3.

R binary files can also be written out and loaded into an R session. These have the advantage of being very efficient at storing data and quicker to load than, for example, .csv or .txt files. The code below saves the data R object – check your working directory when you have run this to see the differences in file size on your computer:

```
save(list = c("df"), file = "df.RData")
```

Multiple R objects can be saved in the same .RData file:

```
save(list = c("df", "tb"), file = "data.RData")
```

The .Rdata files can be opened using the load function:

```
load(file = "data.RData")
```

What this does is load in the R objects from the .RData file to the R session, with the same names. To test this run the code below, if you have run the code snippet above. This deletes two R objects and then loads them back in:

```
ls()
rm(list = c("df", "tb"))
ls()
load(file = "data.RData")
ls()
```

The entire workspace of all R objects can also be saved:

```
save.image(file = "wksp.RData")
```

You can use the read.csv, read_csv, read.table and load functions to read data directly from a URL. The code below loads a .RData file with nine R objects (datasets) and a .CSV file of real estate transactions in the Sacramento area:

```
# load 9 R objects in a single but remote RData file
url=url("http://www.people.fas.harvard.edu/~zhukov/Datasets.RData")
load(url)
ls()
```

```
# read_csv / read.csv
real_estate_url="https://raw.githubusercontent.com/lexcomber/
                 bookdata/main/sac.csv"
real_estate = read_csv(real_estate_url)
real_estate
```

There are also tools for loading data in other formats such as the foreign package. An internet search can quickly provide solutions for loading data in different formats.

In summary, the first observable difference between tibbles and standard data frames is that when a *tibble* is printed out a truncated version is seen, with no need to use head to stop every row from being printed. Also, they state what *kind* of variable each column is when they are printed. Thus, in tb the weight, horse_power and q_mile variables are all indicated as numeric double precision numbers (<dbl>) but type is a character (<chr>) variable. Information about the unprinted rows and columns is also given. Other differences are in the way tibbles are read in from files – for example, they will always reproduce the column names in .csv files as they are given – so that a column with the header (name) of 'Life Exp' would be stored as Life Exp in a tibble rather than Life.Exp in a standard data frame. Also, character variables are stored as such, rather than converted to factors. If a factor variable is genuinely required, an explicit conversion should be used. To read a .csv file into a tibble, read_csv is used in place of read.csv. Perhaps the most important difference is that the *tibble* is surly and complains if operations on it are not correctly specified, whereas the *data frame* is too accommodating – it tries to help – for example in partial matching, which is dangerous in data science.

Key Points

- The tibble format is a reworking of the data.frame format that retains its advantages and eliminates less effective aspects.

- It seeks to be *lazy* and will not link to partially matched variable names (unlike the data.frame format).

- It is *surly* and complains if calls to the data are not correct.

- A number of different ways of reading and writing data from and to external files were illustrated.

- Package *vignettes* were introduced. These provide short overviews of particular aspects of package functionality and operations.

2.2.4 Spatial data formats: sf vs. sp

Data describing spatial features in R are similar in structure to tabular data, but they also include information about the spatial properties of each observation: the coordinates of the point, line or area. These allow the geography of the observation to be interrogated spatially. The structural similarity of spatial data tables to ordinary data tables allows them to be manipulated in much the same way.

For many years the tools for handling spatial data in R were built around the spatial data structures defined in the sp package. The sp format for spatial data in R is a powerful structure that supports a great deal of spatial data manipulation. In this data model, spatial objects can be thought of as being divided into two parts – a data frame (essentially the same as those considered until now) and the geometric information for one of several kinds of spatial objects (e.g. SpatialPointsDataFrame, SpatialPolygonsDataFrame), where each row in the data frame is associated with an individual component of the geometric information. The sp class of objects is broadly analogous to shapefile formats (lines, points, areas) and raster or grid formats. The sp format defines spatial objects both with a data.frame (holding attributes) and without (purely spatial objects) as shown in Table 2.3.

Table 2.3 The sp class of spatial objects

Without attributes	With attributes	ArcGIS equivalent
SpatialPoints	SpatialPointsDataFrame	Point shapefile
SpatialLines	SpatialLinesDataFrame	Line shapefile
SpatialPoygons	SpatialPoygonsDataFrame	Polygon shapefile
SpatialPixels	SpatialPixelsDataFrame	Raster or grid
SpatialGrid	SpatialGridDataFrame	Raster or grid

However, recently a new class of spatial object has been defined called sf (which stands for *simple features*). The sf format seeks to encode spatial data, in a way that conforms to formal standards defined in the ISO 19125-1:2004 standard.

This mode of storage is more 'in tune' with the tidy framework. Essentially spatial objects in sf format appear as a data frame with an extra column named geometry that contains the geometrical information for the spatial part of the object. A column containing a geometric feature is called an sfc (simple feature column), and this can be used in various overlay and other geometric operations, but of itself is not a tibble or data frame. A data frame containing a geometry column of sfc type is called an sf object. It is relatively easy to create both kinds of object via the sf package. Since a great deal of geographical data exists either through existing 'classical' R data objects (from sp) or as shapefiles, a common approach is to convert existing sp objects from a Spatial*DataFrame (using st_as_sf) or to read from a spatial data file such as shapefile or geopackage, using st_read() – both of which are functions provided in the sf package. Thus, sf emphasises the spatial geometry of objects, their hierarchical ordering and the way that objects are stored in databases. The team developing sf (many of whom also developed the sp package and format) aim to provide a new, consistent and standard format for spatial data in R, and the sf format implements a *tidy* data philosophy in the same way as tibble.

Data are tidy (Wickham, 2014; Wickham and Grolemund, 2016) when:

- Each variable forms a column

- Each observation forms a row

- Each type of observational unit forms a table.

Further tidy aspects of the sf package are:

- All functions/methods start with st_ (enter this in the console and press tab to search), use _ and are in lower case

- All functions have the input data as the first argument and are thus piping syntax friendly (see Chapter 3)

- By default stringAsFactors = FALSE (see above describing tibble and data.frame formats)

- dplyr verbs can be directly applied to sf objects, meaning that they can be manipulated in the same way as ordinary data tables such as tibble format (Chapters 3, 4 and 7).

The idea behind sf is that a *feature* (an object in a spatial data layer or a spatial database representing part of the real world) is often composed of other objects, with a set of objects forming a single feature. A forest stand can be a feature, a forest can be a feature, a city can be a feature. Each of these has other objects (features) within it. Similarly, a satellite image pixel can be a feature, but so can a complete image as well. Features in sf have a geometry describing where they are located, and they have attributes which describe their properties. An overview of the evolution of spatial data formats in R can be found at https://edzer.github.io/UseR2017/.

The sf package puts features in tables derived from data.frame or tibble formats. These tables have simple feature class (sfc) geometries in a column, where each element is the geometry of a single feature of class sfg. Feature geometries are represented in R by:

- a numeric vector for a single point (POINT)

- a numeric matrix (each row a point) for a set of points (MULTIPOINT or LINESTRING)

- a list of matrices for a set of set of points (MULTIINESTRING, POLYGON)

- a list of lists of matrices (MULTIPOLYGON)

- a list of anything mentioned above (GEOMETRYCOLLECTION) – all other classes also fall in one of these categories.

To explore this, the code below loads a spatial dataset of polygons representing the counties in North Carolina, USA that is included in the sf package (see https://cran.r-project.org/web/packages/spdep/vignettes/sids.pdf):

```
# read the data
nc <- st_read(system.file("shape/nc.shp", package="sf"), quiet = T)
```

The nc object can be mapped using the qtm() function in the tmap package, a bespoke mapping package (Tennekes, 2018), as in Figure 2.1. Note that tmap can work with both sp and sf data formats.

```
# do a quick tmap of the data
qtm(nc)
```

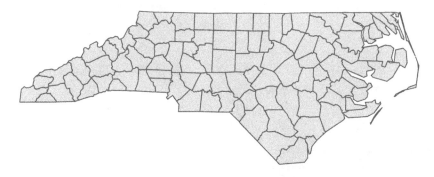

Figure 2.1

However, at the time of writing (and for some years yet) sp formats are still required by many packages for spatial analysis which have not yet been updated to work with sf either indirectly or directly. Many packages still have *dependencies* on sp. An example is the GWmodel package (Lu et al., 2014) for geographically weighted regression (Brunsdon et al., 1996) which at the time of writing still has dependencies on sp, maptools, spData and other packages. If you install and load GWmodel you will see these packages being loaded. However, ultimately sf formats will completely replace sp and packages that use sp will eventually be updated to use sf, but that may be a few years away.

It is easy to convert between **sp** and **sf** formats:

```
# sf to sp
nc_sp <-as(nc, "Spatial")
# sp to sf
nc_sf = st_as_sf(nc_sp)
```

We can examine spatial data in the same way as non-spatial data:

```
dim(nc)
head(nc)
class(nc)
str(nc)
```

You can see that nc has 100 rows and 15 columns and that it is of sf class as well as data.frame. For sf-related formats, R prints out just the first 10 records and all columns of the spatial data table, as with tibble. Try entering:

```
nc
```

You can generate summaries of the nc data table and the attributes it contains:

```
summary(nc)
```

The geometry and attributes of sf objects can be plotted using the plot method defined in the sf and sp packages:

```
# sf format
plot(nc)
plot(st_geometry(nc), col = "red")
# add specific counties to the plot
plot(st_geometry(nc[c(5,6,7,8),]), col = "blue", add = T)
# sp format
plot(nc_sp, col = "indianred")
# add specific counties to the plot
plot(nc_sp[c(5,6,7,8),], col = "dodgerblue", add = T)
```

Spatial data attributes can be mapped using the qtm() function. The code below calculates two rates (actually non-white live births) and then generates Figure 2.2:

```
# calculate rates
nc$NWBIR74_rate = nc$NWBIR74/nc$BIR74
nc$NWBIR79_rate = nc$NWBIR79/nc$BIR79
qtm(nc, fill = c("NWBIR74_rate", "NWBIR79_rate"))
```

The code above starts to hint at the similarities between data tables and spatial data. Recall that the difference between spatial data and ordinary wide data tables is that each observation in a spatial dataset is associated with a location – a point, line or area. The result is that the data tables behind spatial data formats can be interrogated, selected, and so on in much the same way as ordinary data tables, using the same kinds of operations, but the results have a spatial dimension. This is the case for spatial data in both sp and sf formats.

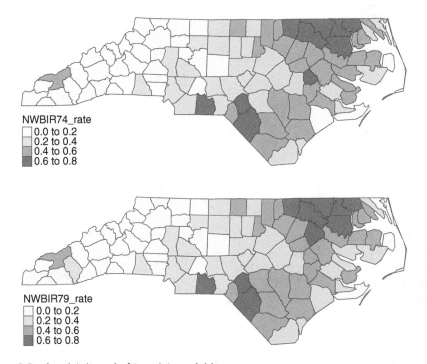

Figure 2.2 A quick 'tmap' of two data variables

The code below uses the nc spatial layer to select specific fields and records for further field manipulations, for spatial operations or for mapping.

- Select and manipulate a *data* field (this returns data tables, not spatial data). Note how the st_drop_geometry function is used for multiple fields to return a data.frame:

```
# look at the field names
names(nc)
# select single field using $
nc$AREA
nc$AREA * 2
```

```
# select multiple fields
st_drop_geometry(nc[,c("AREA", "BIR79", "SID79")])
st_drop_geometry(nc[,c(1,12,13)])
```

- Select records and *data* fields (to return a data table, not spatial data):

```
names(nc)
# select first 50 elements from a single field using $
nc$AREA[1:50]
# randomly select 50 elements from a single field using $
nc$AREA[sample(50, nrow(nc))]
# select first 10 records from multiple fields
st_drop_geometry(nc[1:10 ,c("AREA", "BIR79", "SID79")])
st_drop_geometry(nc[1:10,c(1,12,13)])
```

- Select and manipulate *spatial data* fields and records to return spatial data objects, which can of course be plotted:

```
# selecting spatial data with specific fields
nc[,1]
nc[,"AREA"]
nc[,c(1,12,13)]
nc[, c("AREA", "BIR79", "SID79")]
# selecting specific records and fields
nc[1:6 ,c(1,12,13)]
nc[c(1,5,6,7,8,19), c("AREA", "BIR79", "SID79")]
# plot records 5,6,7,8
plot(st_geometry(nc[c(5,6,7,8),]), col = "red")
```

The sf package has a number of vignettes or tutorials that you should explore. These include an overview of the format, reading and writing from and to sf formats, including conversions to and from sp and sf, and some illustrations of how sf objects can be manipulated. The code below will create a new window with a list of sf vignettes:

```
library(sf)
vignette(package = "sf")
```

To display a specific vignette topic, this can be called using the vignette() function. A full description of the sf format can be found in the sf1 vignette:

```
vignette("sf1", package = "sf")
```

It is possible to save sf objects to R binary files as with any R object:

```
save(list = "nc", file = "nc.RData")
```

R objects in .Rdata files can be loaded to an R session using the load function:

```
load(file = "nc.RData")
```

You may have noticed that the North Carolina dataset was read using the st_read function. Spatial data in sf format can also be written using the st_write() function. For example, to write nc (or any other simple features object) to a spatial data format, at least two arguments are needed: the object and a filename.

The code below writes the nc object to a shapefile format. Note that this will not work if the nc.shp file exists in the working directory, so the delete_layer = T parameter needs to be specified:

```
st_write(nc, "nc.shp", delete_layer = T)
```

The filename is taken as the data source name. The default for the layer name is the basename (filename without path) of the data source name. For this, st_write guesses the driver (for the format of the external spatial data file), which in this case is an ESRI shapefile. The above command is, for instance, equivalent to:

```
st_write(nc, dsn = "nc.shp", layer = "nc.shp",
         driver = "ESRI Shapefile", delete_layer = T)
```

Typical users will use a filename that includes a path, or first set R's working directory with setwd() and use a filename without a path.

Spatial data in sf format can also be written out to a range of different spatial data formats:

```
# as GeoPackage
st_write(nc, "nc.gpkg", delete_layer = T)
```

The formats for vector data are listed at https://gdal.org/drivers/vector/index.html. Generally the format and the output driver are correctly guessed by st_write from the data source name, either from its extension (.shp: ESRI shapefile), or from its prefix (PG: PostgreSQL). The list of extensions with corresponding driver (short driver name) can be found in the sf2 vignette:

```
vignette("sf2", package = "sf")
```

It is also possible to read spatial data into R using the st_read function:

```
new_nc = st_read("nc.shp")
new_nc1
new_nc = st_read("nc.gpkg")
new_nc2
```

Finally, the data table can be extracted from sf and sp objects (i.e. without spatial attributes). For sf objects the geometry can be set to NULL or the st_drop_geometry() function can be used. For sp objects, the data table can be accessed using the data.frame function or the @data syntax. In all cases a data.frame is created and can be examined:

```
# 1. sf format
# create a copy of the nc data
nc2 <- nc
# 1a. remove the geometry by setting to NULL
st_geometry(nc2) <- NULL
class(nc2)
head(nc2)
# 1b. remove geometry using st_drop_geometry
nc2 = st_drop_geometry(nc)
class(nc2)
head(nc2)
# 2. sp format
# 2a. using data.frame
class(data.frame(nc_sp))
head(data.frame(nc_sp))
# 2b. using @data
class(nc_sp@data)
head(nc_sp@data)
```

Key Points

- The *simple features* format for spatial data is defined in the sf package.
- The sf format is gradually replacing the sp format, but some interchange is necessary as some packages still only take sp format as inputs.
- Spatial data tables (in both sf or sp formats) can be interrogated in much the same way as ordinary data tables.
- A number of different ways of reading and writing spatial data from and to external files were illustrated.
- Some basic mapping of spatial data and spatial data attributes was introduced.

2.3 THE TIDYVERSE AND TIDY DATA

A collection of R libraries often referred to as the *tidyverse* was created by Hadley Wickham and associates working at RStudio to provide a more streamlined approach to manipulating and analysing data in R. Two key characteristics of the tidyverse are *tidy data* and *pipelining*.

Recall that the tidy data idea essentially requires (1) a single variable in its own column, (2) each observation in it its own row, and (3) each value in its own cell. An example of a dataset *not* complying with this can be derived from the AirPassengers dataset supplied in the package datasets:

```
AirPassengers
##      Jan Feb Mar Apr May Jun Jul Aug Sep Oct Nov Dec
## 1949 112 118 132 129 121 135 148 148 136 119 104 118
## 1950 115 126 141 135 125 149 170 170 158 133 114 140
## 1951 145 150 178 163 172 178 199 199 184 162 146 166
## 1952 171 180 193 181 183 218 230 242 209 191 172 194
## 1953 196 196 236 235 229 243 264 272 237 211 180 201
## 1954 204 188 235 227 234 264 302 293 259 229 203 229
## 1955 242 233 267 269 270 315 364 347 312 274 237 278
## 1956 284 277 317 313 318 374 413 405 355 306 271 306
## 1957 315 301 356 348 355 422 465 467 404 347 305 336
## 1958 340 318 362 348 363 435 491 505 404 359 310 337
## 1959 360 342 406 396 420 472 548 559 463 407 362 405
## 1960 417 391 419 461 472 535 622 606 508 461 390 432
```

This dataset provides monthly totals of airline passengers from 1949 to 1960. Initially this is in a time series (ts) format, and when printed out takes matrix form, with rows as years, and columns as months. This can be transformed to a data.frame object of the same form:

```
AirPassengers_df <- matrix(datasets::AirPassengers,12,12,
                           byrow = TRUE)
rownames(AirPassengers_df) <- 1949:1960
colnames(AirPassengers_df) <- c("Jan","Feb","Mar","Apr","May", "Jun",
                                "Jul","Aug","Sep","Oct","Nov","Dec")
AirPassengers_df <- data.frame(AirPassengers_df)
head(AirPassengers_df)
##      Jan Feb Mar Apr May Jun Jul Aug Sep Oct Nov Dec
## 1949 112 118 132 129 121 135 148 148 136 119 104 118
## 1950 115 126 141 135 125 149 170 170 158 133 114 140
## 1951 145 150 178 163 172 178 199 199 184 162 146 166
## 1952 171 180 193 181 183 218 230 242 209 191 172 194
## 1953 196 196 236 235 229 243 264 272 237 211 180 201
## 1954 204 188 235 227 234 264 302 293 259 229 203 229
```

AirPassengers_df prints out in the same way as AirPassengers but the two variables are stored differently internally by R. We will consider AirPassengers_df as an example of non-tidy data – but also as a plausible format for, for example, sharing this information.

A typical way of modelling data in this way might be to consider a monthly passenger count as an observation, which could be seen as a function of the year (trend over time) and month (seasonal fluctuation). Month is a variable, but is spread over *several* columns in the format used in AirPassengers_df. Thus requirement (1) above is violated. In addition, since several observations are on the same row, requirement (2) is also violated.

Of course, these 'violations' are only considered as such in terms of the tidyverse – it is possible to manipulate, analyse and visualise data in this format. However, it is argued that having a multiplicity of data formats makes it harder to communicate how information is organised and to reproduce results. To convert these data into a tidy format, gather may be used.

```
AirPassengers_df$Year <- rownames(AirPassengers_df)
AirPassengers_tidy <- gather(AirPassengers_df, key=Month,
                                  value=Count,-Year)
head(AirPassengers_tidy)

##     Year Month  Count
## 1 1949    Jan     112
## 2 1950    Jan     115
## 3 1951    Jan     145
## 4 1952    Jan     171
## 5 1953    Jan     196
## 6 1954    Jan     204
```

We will consider the details of the above code in Chapter 3, but for now note that the data frame AirPassengers_tidy *is* tidy. Each observation is a row: a single row contains the information of a unique passenger count, together with the month and the year it refers to. Similarly, the three variables Count, Month and Year each have a unique column.

Note also that in this format, the order of the rows does not affect the content of the data, in contrast to the original AirPassengers data which implicitly implied that counts were on a month-by-month basis starting from January 1949.

A final point here is that the year was stored in the rownames attribute of the data frame (again, the tidy principle suggests this should be a standard variable), here called Year. Again, *all* of the variables have a column – but it should be an ordinary column to avoid confusion, rather than be 'hidden' in a different kind of attribute called rowname.

Of course, it cannot be guaranteed that all the data that a data scientist receives will be tidy. However, a key aspect of the tidyverse in terms of data manipulation and 'wrangling' is that it provides a suite of functions that take tidy data as input, and return transformed but still tidy data as output. To complement this, there are also tools to take non-tidy data and restructure them in a tidy way. Some of these come with base R (e.g. `AirPassengers_df$Year <- rownames(AirPassengers_df)` ensures the row names are stored as a standard variable), and others are supplied as part of the tidyverse (e.g. `gather()`). A standard workflow here is as follows:

1. Make data tidy if they are not so already.

2. Manipulate the data through a series of 'tidy-to-tidy' transformations.

Key Points

- Tidy data are defined as data in which a single variable has its own column, each observation is in its own row. and each value has its own cell.

- Untidy data can be converted into tidy format using the `gather()` function.

2.4 `dplyr` FOR MANIPULATING DATA (WITHOUT PIPES)

2.4.1 Introduction to `dplyr`

The `dplyr` package is part of `tidyverse` and provides a suite of tools for manipulating and summarising data tables and spatial data. It contains a number of functions for data transformations that take a tidy data frame and return a transformed tidy data frame. A very basic example of this is `filter()`: this simply selects a subset of a data frame on the basis of its rows. Given the requirement that each row corresponds to a unique observation, this effectively subsets observations.

We can use the air passenger data, for example, in order to find the subset of the data for the passenger counts in March:

```
head(filter(AirPassengers_tidy, Month == "Mar"))

##      Year Month  Count
## 1 1949    Mar    132
## 2 1950    Mar    141
## 3 1951    Mar    178
## 4 1952    Mar    193
## 5 1953    Mar    236
## 6 1954    Mar    235
```

A few things to note here:

1. The first argument of `filter` is the tidy data frame.

2. The second argument is a logical expression specifying the subset.

3. Column names in the data frame can be used in the logical expression more or less as though they were ordinary variables.

4. The output, as stated, is a tidy data frame as well – in the same format as the input, but with fewer observations (only those whose month is March).

The logical expression can be any valid logical expression in R – thus >, <, <=, >=, == and != may all be used, as can & (and), | (or) and ! (not). The & operator is not always necessary, as it is possible to apply multiple conditions in the same filter operation by adding further logical arguments:

```
filter(AirPassengers_tidy,Month=="Mar",Year > 1955)

##    Year Month  Count
## 1 1956    Mar    317
## 2 1957    Mar    356
## 3 1958    Mar    362
## 4 1959    Mar    406
## 5 1960    Mar    419
```

Another data transform offered by dplyr is arrange(), which sorts the data in the order of one of the variables. As was mentioned earlier, the data frame should contain all of the information needed regardless of order, but as a matter of presentation sometimes it is useful to see the data in some particular order.

```
head(arrange(AirPassengers_tidy,Year))

##    Year Month  Count
## 1 1949    Jan    112
## 2 1949    Feb    118
## 3 1949    Mar    132
## 4 1949    Apr    129
## 5 1949    May    121
## 6 1949    Jun    135
```

To sort in *descending* order use the desc() function:

```
head(arrange(AirPassengers_tidy,desc(Year)))
```

For datasets with several columns, several sorting variables can be applied, and the sort will be nested within variables, running from left to right. This is demonstrated using the data frame state.x77 also supplied in the package datasets – also linking to the variables state.division and state.region which provide broader geographical locations for each US state. To make it tidy the following R code is used:

```
state_tidy <- as.data.frame(state.x77)
state_tidy$Division <- as.character(state.division)
state_tidy$Region <- as.character(state.region)
state_tidy$Name <- rownames(state.x77)
```

The output can be examined:

```
head(state_tidy)
```

This turns state.x77 from a matrix into a data frame, adds state.region and state.division as columns (of data type character) and makes the row names (actually the names of the states) into a *bona fide* column (they were the data frame row names in state.x77). Now the data may be sorted alphabetically by region, and within that by division, and finally within that by life expectancy:

```
head(arrange(state_tidy,Region,Division,`Life Exp`))
##   Population  Income Illiteracy Life Exp Murder HS Grad
## 1     11197    5107        0.9    70.14   10.3    52.6
## 2      9111    4751        0.9    70.63   11.1    52.8
## 3     10735    4561        0.8    70.82    7.4    53.2
## 4      5313    4458        0.7    70.88    7.1    52.9
## 5      4589    4468        0.7    72.48    3.0    54.5
## 6      4767    4254        0.8    70.69    9.3    48.8
##   Frost  Area           Division          Region      Name
## 1   127 55748 East North Central North Central   Illinois
## 2   125 56817 East North Central North Central   Michigan
## 3   124 40975 East North Central North Central       Ohio
## 4   122 36097 East North Central North Central    Indiana
## 5   149 54464 East North Central North Central  Wisconsin
## 6   108 68995 West North Central North Central   Missouri
```

Because the life expectancy variable name Life Exp contains a space, we need to enclose it in backward quotes when referring to it. Again this transformation has the *tidy-to-tidy* pattern so that an additional filter command could be applied to the output. Having arranged the state data as above, we may want to filter out the places where average life expectancy falls below 70 years:

```
filter(arrange(state_tidy,Region,Division,`Life Exp`),
       `Life Exp`<70)
```

Key Points

- The dplyr package was introduced.

- Some basic `dplyr` operations were illustrated, including `filter` to select rows and `arrange` to order rows.

- The syntax for `dplyr` functions is standardised with the first argument always the input data. This is important for piping (see Chapter 3).

2.4.2 Single-table manipulations: `dplyr` verbs

The `filter()` and `arrange()` functions are just two of the `dplyr` *verbs* (the package developers call them 'verbs' because they *do* something!). These can be used to manipulate data singly or in a nested sequence of operations with intermediate outputs. Chapter 3 shows that these `dplyr` tools can be used in a piping syntax that chains sequences of data operations together in a single data manipulation workflow. The important `dyplyr` verbs are summarised in Table 2.4. The code snippets below illustrate how some of the single-table `dplyr` verbs can be applied and combined in a *non-piped* way.

Table 2.4   The single-table verbs for manipulating data in `dplyr`

Verb	Description
`filter()`	Selects a subset of rows in a data frame according to user-defined conditional statements
`slice()`	Selects a subset of rows in a data frame by their position (row number)
`arrange()`	Reorders the row order according to the columns specified (by 1st, 2nd and then 3rd column, etc.)
`desc()`	Orders a column in descending order
`select()`	Selects the subset of specified columns and reorders them vertically
`distinct()`	Finds unique values in a table
`mutate()`	Creates and adds new columns based on operations applied to existing columns (e.g. NewCol = Col1 + Col2)
`transmute()`	As `select()` but only retains the new variables
`summarise()`	Summarises values with functions that are passed to it
`sample_n()`	Takes a random sample of table rows
`sample_frac()`	Selects a fixed fraction of rows

The `dplyr` verbs can be used to undertake a number of different operations on the dataset to subset and fold it different ways. We have already used `filter()` to select records that fulfil some kind of logical criterion and `arrange()` to reorder them. The code below uses `mutate()` to create a population density attribute (pop_dens) and assigns the result to a temporary file (tmp) which can be examined:

```
tmp = mutate(state_tidy, pop_dens = Population/Area)
head(tmp)
```

Such manipulations can be combined with other dplyr *verbs*. The code below calculates population density, creating a variable called **pop_dens**. It then selects this and the Name variables and assigns the result to **tmp**:

```
tmp = select(
    mutate(state_tidy, pop_dens = Population/Area),
    Name, pop_dens)
head(tmp)
```

It is possible to group the analyses, for example by division (region), and to generate some summary values over the grouping using the summarise() function (this returns a **tibble** of nine records and four fields):

```
summarise(
    group_by(state_tidy, Division),
    mean_PopD = mean(Illiteracy),
    mean_Income = mean(Income),
    mean_HSG = mean(`HS Grad`)
    )
```

It is possible to further combine verbs by nesting them. The code below nests a mutate() to create a new variable within a group_by() and summarise() combination to determine group means that are then ordered by arrange():

```
tmp = arrange(
        summarise(
            group_by(
                mutate(state_tidy, Inc_Illit = Income/Illiteracy),
            Division
            ),
            mean_II = mean(Inc_Illit)),
        mean_II)
```

You can inspect the results in tmp. But notice how complex this is getting – keeping track of the nested statement and the parentheses!

It would be preferable *not* to have to create intermediary variables like tmp for longer chains of analysis. This is especially the case if we want to do something like pass the results to ggplot so that they can be visualised, such as in an ordered lollipop plot as in Figure 2.3:

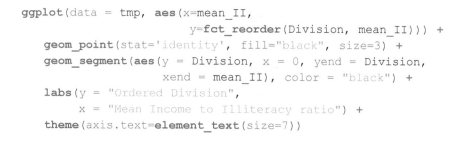

```
ggplot(data = tmp, aes(x=mean_II,
                        y=fct_reorder(Division, mean_II))) +
    geom_point(stat='identity', fill="black", size=3) +
    geom_segment(aes(y = Division, x = 0, yend = Division,
                     xend = mean_II), color = "black") +
    labs(y = "Ordered Division",
         x = "Mean Income to Illiteracy ratio") +
    theme(axis.text=element_text(size=7))
```

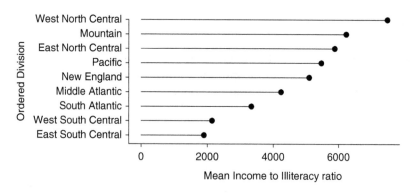

Figure 2.3   Mean income to illiteracy ratios for the US divisions

This is the first time that you have seen the ggplot() syntax. The ggplot2 package contains an almost infinite number of ways of visualising data. It is given a comprehensive treatment in Chapter 5. The basic ggplot() syntax requires two components to be specified:

(a) the plot initialisation that specifies the data and what are called the *mapping aesthetics* (aes()), or the variables to be mapped with the ggplot() function; and

(b) the plot type itself – in this case (the lollipop plot in Figure 2.3) the main plot is the geom_point() which plots the dots, but this is supplemented by a call to geom_segment() which plots the lines.

It is important to note that each set function used in ggplot() is joined by a + sign. In the construction of Figure 2.3 these also include some labelling instructions and specify label text size. These hint at the theoretical underpinnings of ggplot() which are from Wickham (2010). In this each component of the plot is considered as a layer, hence the need for the + between each of the plotting instructions.

Back to dplyr. Chapter 3 describes how to create chains of dplyr data manipulations using what is called *piping syntax*. In this the results of one operation are directly passed or piped to another operation. They can also be piped directly to ggplot(). The code snippet below undertakes the same operations as the two snippets above, without creating any intermediary variables (tmp). This is the first time you have seen piping syntax in this book, and the aim here is to simply intro-duce it. The pipe is denoted by %>% and it pushes the output of one operation into the input of another. The rest of this chapter has some piping and this will serve as an introduction to the next section and provide a grounding for the comprehensive treatment of piping syntax in Chapter 3.

```r
state_tidy %>%
  # create Inc_Illit
  mutate(Inc_Illit = Income/Illiteracy) %>%
  # group by Division
  group_by(Division) %>%
  # calculate group summaries
  summarise(mean_II = mean(Inc_Illit)) %>%
  # pass to ggplot
  ggplot(aes(x=mean_II, y=fct_reorder(Division, mean_II))) +
  geom_point(stat='identity', fill="black", size=3) +
  geom_segment(aes(y=Division, x = 0,yend=Division,
                   xend=mean_II),  color = "black") +
  labs(y="Ordered Division",x="Mean Income to Illiteracy ratio") +
  theme(axis.text=element_text(size=7))
```

The above code may appear confusing, but each line before the line with ggplot() can be run to see what the code is doing with the data. The aim here is to stress the importance of understanding the operation of these verbs as they underpin piping and data analytics with databases described in Chapter 4. So, for example, try examining the following sequence of code snippets:

```r
state_tidy %>%
  mutate(Inc_Illit = Income/Illiteracy)
```

Then…

```r
state_tidy %>%
  mutate(Inc_Illit = Income/Illiteracy) %>%
  group_by(Division)
```

And finally…

```r
state_tidy %>%
  mutate(Inc_Illit = Income/Illiteracy) %>%
  group_by(Division) %>%
  summarise(mean_II = mean(Inc_Illit))
```

It is also important to note that the `dplyr` functions can be applied to `sf` format spatial objects. The code below calculates a non-white birth rate from the NWB74 and NWBIR74 variables in the `nc` spatial layer, and overwrites `nc` with the new `sf` object:

```
nc = mutate(nc, Rate = NWBIR74 / BIR74)
qtm(nc, "Rate", fill.palette = "Greens")
```

The different `dplyr` single-table verbs are used extensively throughout this book. You will find that your use of the `dplyr` functions for manipulating tables will grow in complexity the more you use them, and especially with the piping syntax. The key consideration as you progress with your data manipulations using `dyplyr` is the order or sequence of operations.

You should explore the introductory vignette to `dplyr` which describes the single-table verbs in `dplyr`:

```
vignette("dplyr", package = "dplyr")
```

Key Points

- The `dplyr` package has a number of *verbs* for data table manipulations.
- These can be used to *select* variables or fields and to *filter* records or observations using logical statements.
- They can be used to create new variables using the `mutate()` function.
- Group summaries can be easily generated using a combination of the `group_by()` and `summarise()` functions.
- Sequences of *nested* `dplyr` verbs can be applied in a code block.
- Many `dplyr` operations can be undertaken on `sf` format spatial data.

2.4.3 Joining data tables in `dplyr`

Tables can be joined through an attribute that they have in common. To illustrate this, the code below loads some census data for Liverpool held in an R binary file (`ch2.Rdata`) and saves it to your current working directory – you may want to clear your workspace to check this:

```
rm(list = ls())
getwd()
download.file("http://archive.researchdata.leeds.ac.uk/731/1/
                ch2.RData",
              "./ch2.RData", mode = "wb")
```

```
load("ch2.RData")
ls()

## [1] "lsoa"   "lsoa_data"
```

You will notice that two objects lsoa and lsoa_data are loaded into your R session (enter ls()). The lsoa object is a spatial layer of Lower Super Output Areas for Liverpool in sf format with 298 areas, and lsoa_data is a tibble with 298 rows and 11 columns.

The design of the UK Census reporting areas aimed to provide a consistent spatial unit for reporting population data at different scales. Lower Super Output Areas (Martin, 1997, 1998, 2000, 2002) contain around 1500 people. This means that their size is a function of population density. UK Census data can be downloaded from the Office of National Statistics (https://www.nomisweb.co.uk) and spatial layers of Census areas from the UK Data Service (https://borders.ukdata service.ac.uk).

You could examine the tibble and sf objects that have been loaded:

```
lsoa_data
lsoa
plot(st_geometry(lsoa), col = "thistle")
```

Both have a field called code and this can be used to link the two datasets using one of the dplyr join functions:

```
lsoa_join = inner_join(lsoa, lsoa_data)
```

There are a few things to note here:

- By default dplyr joins look for similarly named fields to try to join them by, and in this case both inputs have a field called code.

- Note also that the join function returns an object in the same format as the first argument, in this case an sf spatial layer. You could try reversing the order in the snippet above to examine this.

- Finally, a warning tells us that the join has been made but that a factor and character format variable have been joined. To overcome this, the attributes can be converted to character, with the result that the join command does not return the warning:

```
lsoa$code = as.character(lsoa$code)
inner_join(lsoa, lsoa_data)
```

It is also possible to specify join fields in situations where the names are not the same. To illustrate this the code below changes the name of code in lsoa_data, specifies the join and then renames the field back to code:

```
names(lsoa_data)[1] = "ID"
lsoa_join = inner_join(lsoa, lsoa_data, by = c("code" = "ID"))
names(lsoa_data)[1] = "code"
```

It is important to note that links between two tables can be defined in different ways. The different join types available in dplyr are summarised in Table 2.5. There are six join types included in the dplyr package – four of them are what are called *mutating* joins and two *filtering* joins. Mutating joins combine variables from both the x and y inputs and filtering joins only keep records from the x input. They have a similar syntax:

```
result <- JOIN_TYPE(x, y)
```

Table 2.5 The different types of join in the dplyr package

Join	Description	Type
inner_join()	Returns all of the records from x with matching values in y, and all fields from both x and y. If there are multiple matches between x and y, all combination of the matches are returned.	mutating
left_join()	Returns all of the records from x, and all of the fields from x and y. Any records in x with no match in y will be given NA values in the new fields. If there are multiple matches between x and y, all combination of the matches are returned.	mutating
right_join()	Returns all of the records in y, and all of the fields in x and y. Records in y with no match in x are given NA values in the new fields. If there are multiple matches between x and y, all combination of the matches are returned.	mutating
full_join()	Returns all records and fields from both x and y. NA is allocated to any non-matching values.	mutating
semi_join()	Returns all records from x where there are matching values in y, keeping just the fields from x. It never duplicates records as does inner_join().	filtering
anti_join()	Returns all the records from x where there are non-matching values in y, keeping just fields from x.	filtering

It is instructive to see how they work with lsoa_data and lsoa, using some mismatched data. The code below randomly samples 250 records from lsoa_data and the inner join means that only records that match are returned, and these are passed to dim to get the object dimensions:

```
set.seed(1984)
lsoa_mis = sample_n(lsoa_data, 250)
# nest the join inside dim
dim(inner_join(lsoa_mis, lsoa))
## [1] 250 12
```

The different join types can return different results. The code snippets below illustrate this using dim():

```
dim(inner_join(lsoa, lsoa_mis))
dim(left_join(lsoa, lsoa_mis))
dim(right_join(lsoa, lsoa_mis))
dim(semi_join(lsoa, lsoa_mis))
dim(anti_join(lsoa, lsoa_mis))
```

You should compare the results when x and y are changed, especially for left and right joins. The links between two tables can be defined in different ways and cardinality enforced between them through the choice of join method. In a perfect world all joins would result in a seamless, single match between each common object in each table, with no unmatched records and no ambiguous matches. Most cases are not like this and the different join types essentially enforce rules for how unmatched records are treated. This is how dplyr enforces *cardinality*, or relations based on the uniqueness of the values in data column being joined. In this context, the mutating joins combine variables from both x and y inputs, filtering the results to include all of the records from the input that is given prominence in the join. So only matching records in x and y are returned with an inner join, all of the elements of y are returned for a right join, etc.).

The help section for dplyr joins has some other worked examples that you can copy and paste into your console. To see these, run:

```
?inner_join
```

You should also work through the vignette for dplyr joins:

```
vignette("two-table", package = "dplyr")
```

As with the dplyr single-table *verbs*, various joins are used extensively throughout this book.

Putting it all together, single-table manipulations and joins can be combined (in fact, they usually *are* combined). The code below mixes some single-table verbs and joining operations to create an unemployment rate, selects the areas within the top 25% and passes this to qtm:

```
qtm (
  filter (
    mutate (
      inner_join(lsoa, lsoa_data),
        unemp_pc = (econ_active-employed)/emp_pop),
      unemp_pc > quantile(unemp_pc, 0.75)), "tomato")
```

This is getting complex. Chapter 3 describes how sequences or chains of dplyr functions can be constructed using a piping syntax, replacing the code above with the piped code below:

```
lsoa %>% inner_join(lsoa_data) %>%
  mutate(unemp_pc = (econ_active-employed)/emp_pop) %>%
  filter(unemp_pc > quantile(unemp_pc, 0.75)) %>%
  qtm("tomato")
      mutate(unemp_pc = (econ_active-employed)/emp_pop) %>%
      filter(unemp_pc > quantile(unemp_pc, 0.75)) %>%
      qtm("tomato")
```

Key Points

- Data tables can be joined through an attribute that they have in common.

- This can be done in different directions, preserving the inputs in different ways (compare left_join with right_join and inner_join).

- Spatial data tables in sf format can be joined in the same way, preserving the spatial properties.

2.5 MAPPING AND VISUALISING SPATIAL PROPERTIES WITH tmap

Having introduced spatial data and done some mapping in this chapter using the qtm() function (qtm generates a *quick tmap*), much greater control over the mapping of spatial data can be exercised using the *full* tmap. This section briefly outlines the tmap package for creating maps from spatial data, providing information on the structure and syntax of calls to tmap. A more comprehensive and detailed treatment on tmap can be found in Brunsdon and Comber (2018).

The tmap package supports the *thematic visualisation* of spatial data (Tennekes, 2018). It has a grammatical style that handles each element of the map separately in a series of layers (it is similar to ggplot in this respect – see Chapter 5). In so doing it seeks to exercise control over each element in the map. This is different from the basic R plot functions.

The basic syntax of tmap is:

```
tm_shape(data = <data>) +
  tm_<function>()
```

Do not run the code above; it simply shows the syntax of tmap. However, you have probably noticed that it uses a similar syntactical style to ggplot to join together layers of commands, using the + sign. The tm_shape() function initialises the map and then layer types, the variables that get mapped, and so on are specified using different flavours of tm_<function>. The main types are listed in Table 2.6.

Table 2.6   Commonly used 'tmap' functions

Function	Description
tm_dots, tm_bubbles	For points either simple or sized according to a variable
tm_lines	For line features
tm_fill, tm_borders, tm_polygons	For polygons/areas, with and without attributes and polygon outlines
tm_text	For adding text to map features
tm_format, tm_facet, tm_layout	For adjusting cartographic appearance and map style, including legends
tm_view, tm_compass, tm_credits, tm_scale_bar	For including traditional cartography elements

Let us start with a simple choropleth map. These maps show the distribution of a continuous variable in different elements of the spatial data (typically polygons/areas or points). The code below creates new variables for percentage unemployed, percentage under 25 and percentage over 65:

```
lsoa_join$UnempPC = (lsoa_join$unemployed/ lsoa_join$econ_active)*100
lsoa_join$u25PC = (lsoa_join$u25/lsoa_join$age_pop)*100
lsoa_join$o65PC = (lsoa_join$o65/lsoa_join$age_pop)*100
```

These can be individually mapped as in Figure 2.4:

```
tm_shape(lsoa_join) +
  tm_polygons("UnempPC")
```

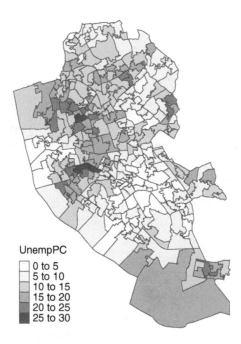

Figure 2.4 A choropleth map of UnempPC

By default tmap picks a shading scheme, the class breaks and places a legend somewhere. All of these can be changed. The code below allocates the tmap plot to p1 (plot 1) and then prints it:

```
p1 = tm_shape(lsoa_join) +
  tm_polygons("UnempPC", palette = "GnBu", border.col = "salmon",
    breaks = seq(0,35, 5), title = "% Unemployed") +
  tm_layout(legend.position = c("right", "top"), legend.outside = T)
p1
```

And of course many other elements can be included either by running the code snippet defining p1 above with additional lines **or** by simply *adding* them as in the code below:

```
p1 + tm_scale_bar(position = c("left", "bottom")) +
  tm_compass(position = c(0.1, 0.1))
```

It is also possible to add or overlay other data such as a boundary, which because this is another spatial data layer needs to be added with tm_shape followed by the usual function:

```
boundary = st_union(lsoa)
p1 + tm_shape(boundary) + tm_borders(col = "black", lwd = 2)
```

The tmap package can be used to plot multiple attributes in the same plot, in this case unemployment and over 65 percentages (Figure 2.5):

```
tm_shape(lsoa_join) +
    tm_fill(c("UnempPC", "o65PC"), palette = "YlGnBu",
        breaks = seq(0,40, 5), title = c("% Unemp", "% Over 65")) +
    tm_layout(legend.position = c("left", "bottom"))
```

Figure 2.5 Choropleth maps of UnempPC and o65PC

The code below applies further refinements to the choropleth map to generate Figure 2.6. Notice the use of title and legend.hist, and then subsequent parameters passed to tm_layout to control the legend:

```
tm_shape(lsoa_join) +
    tm_polygons("u25PC", title = "% Under 25", palette = "Reds",
        style = "kmeans", legend.hist = T) +
    tm_layout(title = "Under 25s in \nLiverpool",
                frame = F, legend.outside = T,
                legend.hist.width = 1,
                legend.format = list(digits = 1),
                legend.outside.position = c("left", "top"),
                legend.text.size = 0.7,
                legend.title.size = 1) +
```

```
tm_compass(position = c(0.1, "0.1")) +
tm_scale_bar(position = c("left", "bottom")) +
tm_shape(boundary) + tm_borders(col = "black", lwd = 2)
```

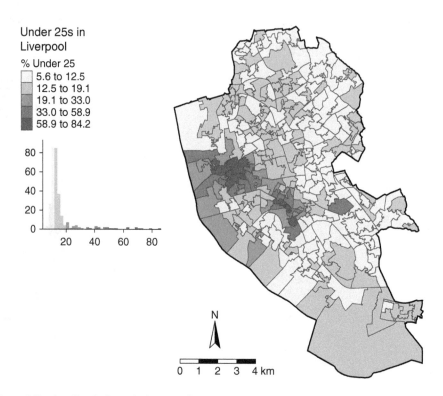

Figure 2.6 A refined choropleth map of u25PC

This is quite a lot of code to unpick. The first call to tm_shape() determines which layer is to be mapped, then a specific mapping function is applied to that layer (in this case tm_polygons()) and a variable is passed to it. Other parameters to specify are the palette, whether a histogram is to be displayed and the type of breaks to be imposed (here a k-means was applied). The help for tm_polygons() describes a number of different parameters that can be passed to tmap elements. Next, some adjustments are made to the defaults through the tm_layout function, which allows you to override the defaults that tmap automatically assigns (shading scheme, legend location, etc.). Finally, the boundary layer is added.

There are literally thousands of options with tmap and many of them are controlled in the tm_layout() function. You should inspect this:

```
?tm_layout
```

The code below creates and maps point data values in different ways. There are two basic options: change the size or change the shading of the points according to the attribute value. The code snippets below illustrate these approaches.

Create the point layer:

```
lsoa_pts = st_centroid(lsoa_join)
```

Then map by size using tm_bubbles() as in Figure 2.7:

```
# the 1st layer
tm_shape(boundary) +
    tm_fill("olivedrab4") +
    tm_borders("grey", lwd = 2) +
    # the points layer
    tm_shape(lsoa_pts) +
    tm_bubbles("o65PC", title.size = "% over 65s", scale = 0.8,
               col = "gold")
```

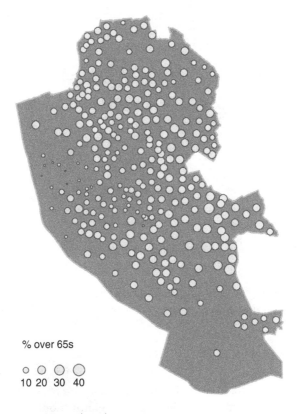

% over 65s

○ ○ ◯ ◯
10 20 30 40

Figure 2.7 Point data values using size

Alternatively, map by shade using `tm_dots` as in Figure 2.8:

```
# the 1st layer
tm_shape(boundary) +
  tm_fill("grey70") +
  tm_borders(lwd = 2) +
  # the points layer
  tm_shape(lsoa_pts) +
  tm_dots("o65PC", size = 0.2, title = "% over 65s", shape = 19) +
  tm_layout(legend.outside = T, frame = F)
```

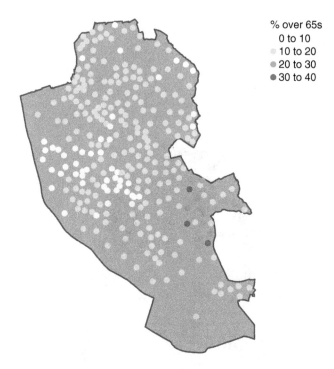

% over 65s
0 to 10
10 to 20
20 to 30
30 to 40

Figure 2.8 Point data values using choropleth mapping (colour)

You should explore the introductory `tmap` vignette which describes other approaches for mapping different types of variables, in-built styles and plotting multiple layers:

```
vignette("tmap-getstarted", package = "tmap")
```

It is possible to use an OpenStreetMap (OSM) backdrop for some basic web mapping with `tmap` by switching the view of the default mode from `plot` to `view`.

You need to be online and connected to the internet. The code below generates the zoomable web map with an OSM backdrop in Figure 2.9, with a transparency term to aid the understanding of the local map context:

```
tmap_mode("view")
tm_shape(lsoa_join) +
    tm_fill(c("UnempPC"), palette = "Reds", alpha = 0.8) +
    tm_basemap('OpenStreetMap')
```

Aerial imagery and satellite imagery (depending on the scale and extent of the scene) can also be used to provide context as in Figure 2.10:

```
tm_shape(lsoa_join) +
    tm_fill(c("UnempPC"), palette = "Reds", alpha = 0.8) +
    tm_basemap('Esri.WorldImagery')
```

Other backdrops can be specified as well as OSM. Try the following: Stamen. Watercolor, Esri.WorldGrayCanvas and Esri.WorldTopoMap.

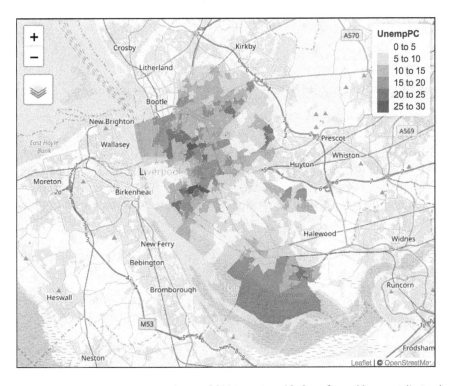

Figure 2.9 Unemployment rates, with an OSM backdrop (© OpenStreetMap contributors)

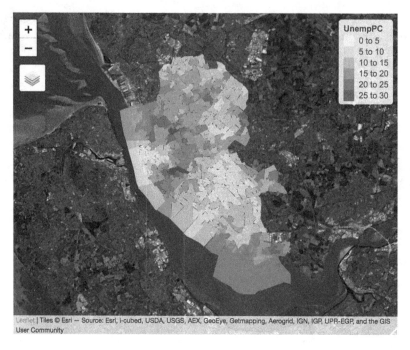

Figure 2.10   Unemployment rates, with an imagery backdrop (© ESRI)

The `tmap_mode` needs to be reset to return to a normal map view:

```
tmap_mode("plot")
```

Finally in this brief description of `tmap`, when maps are assigned to a named R object (like `p1`) you can exert greater control over how multiple maps are plotted together using the `tmap_arrange()` function. This allows the parameters for each map to be specified individually before they are displayed together as in Figure 2.11.

```
p1 <- tm_shape(lsoa_join) +
   tm_fill("UnempPC", palette = "Reds", style = "quantile",
          n = 5, title = "% Unemployed") +
   tm_layout(legend.position = c("left", "bottom")) +
   tm_shape(boundary) + tm_borders(col = "black", lwd = 2)
p2 <- tm_shape(lsoa_join) +
   tm_fill("o65PC", palette = "YlGn", style = "quantile",
          n = 5, title = "% Over 65") +
   tm_layout(legend.position = c("left", "bottom")) +
   tm_shape(boundary) + tm_borders(col = "black", lwd = 2)
```

```
p3 <- tm_shape(lsoa_join) +
   tm_fill("u25PC", palette = "YlOrRd", style = "quantile",
           n = 5, title = "% Under 25") +
   tm_layout(legend.position = c("left", "bottom")) +
   tm_shape(boundary) + tm_borders(col = "black", lwd = 2)
tmap_arrange(p1,p2,p3, nrow = 1)
```

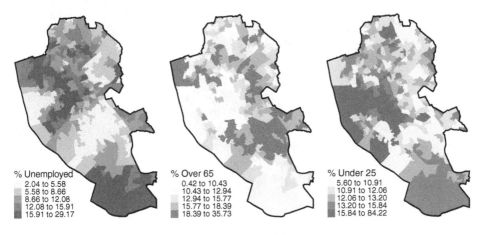

Figure 2.11   The result of combining multiple tmap objects

The full functionality of tmap has only been touched on. It is covered in much greater depth in Brunsdon and Comber (2018), especially in Chapters 3 and 5.

Key Points

- The basic syntax for mapping using the tmap package was introduced.

- This has a layered form tm_shape(data = <data>) + tm_<function>().

- By default tmap allocates a colour scheme, decides the breaks (divisions) for the attribute being mapped, places the legend somewhere, gives a legend title, etc. All of these can be changed.

- Additional spatial data layers can be added to the map.

- tmap has all the usual map embellishments (scale bar, north arrow, etc.).

- tmap allows interactive maps to be created with different map backdrops (e.g. OSM, aerial imagery).

- The tmap_arrange() function allows multiple maps to be plotted together in the same plot window.

2.6 SUMMARY

This chapter has introduced data and spatial data. It has described the characteristics of *wide* and *flat* data formats and shown how to move between them using functions from the `tidyr` package. It described in detail the different characteristics of `tibble` and `data.frame` wide data table formats and of `sf` and `sp` data formats, and how to convert data and spatial data between them. It also provided examples of how to read and write data and spatial into and out of R sessions. It described the characteristics of *tidy* data and introduced some of the principles behind the `tidyverse` and `tibble` data formats.

There are many possible ways to slice, dice and select different elements from data tables. The `dyplyr` package provides a suite of tools for such *data wrangling* and *data munging* – reshaping data into a format that will be helpful for the analysis. A small subset of these have been introduced and illustrated in this chapter, and they are used in many different and more complex ways throughout this book, particularly as they are chained together using piping syntax where the result of one manipulation is passed to another (and then to another…).

REFERENCES

Brunsdon, C. and Comber, L. (2018) *An Introduction to R for Spatial Analysis and Mapping* (2nd edn). London: Sage.

Brunsdon, C., Fotheringham, A. S. and Charlton, M. E. (1996) Geographically weighted regression: A method for exploring spatial nonstationarity. *Geographical Analysis*, 28(4), 281–298.

Comber, A. and Wulder, M. (2019) Considering spatiotemporal processes in big data analysis: Insights from remote sensing of land cover and land use. *Transactions in GIS*, 23(5), 879–891.

Lu, B., Harris, P., Charlton, M. and Brunsdon, C. (2014) The GWmodel R package: Further topics for exploring spatial heterogeneity using geographically weighted models. *Geo-spatial Information Science*, 17(2), 85–101.

Martin, D. (1997) From enumeration districts to output areas: Experiments in the automated creation of a census output geography. *Population Trends*, 88, 36–42.

Martin, D. (1998) 2001 Census output areas: From concept to prototype. *Population Trends*, 94, 19–24.

Martin, D. (2000) Towards the geographies of the 2001 UK Census of Population. *Transactions of the Institute of British Geographers*, 25(3), 321–332.

Martin, D. (2002) Geography for the 2001 Census in England and Wales. *Population Trends*, 108, 7–15.

Tennekes, M. (2018) tmap: Thematic maps in R. *Journal of Statistical Software*, 84(6), 1–39.

Wickham, H. (2010) A layered grammar of graphics. *Journal of Computational and Graphical Statistics*, 19(1), 3–28.

Wickham, H. (2014) Tidy data. *Journal of Statistical Software*, 59(10), 1–23.

Wickham, H. and Grolemund, G. (2016) *R for Data Science: Import, Tidy, Transform, Visualize, and Model Data*. Sebastopol, CA: O'Reilly Media.

A FRAMEWORK FOR PROCESSING DATA: THE PIPING SYNTAX AND dplyr

3.1 OVERVIEW

The previous chapter described the key characteristics for data in the tidyverse to be tidy data, with each variable having its own column, each observation having its own row, and each value having its own cell. Different methods for tidying data were illustrated and the following tidyverse workflow was suggested: make data tidy if they are not so already, then manipulate the data using 'tidy-to-tidy' transformations.

It also introduced a number of dplyr functions for manipulating data, and as part of that process introduced the idea of piping. Recall that piping allows a sequence of data manipulations to be applied. The key thing about this was that there was no need for intermediary or temporary objects to be defined, to hold the results of each step in the analysis or manipulation before passing them to the next step. In a piped sequence of dplyr functions, the result is only returned to the console (and therefore memory) when the pipe ends. In later chapters (e.g. Chapters 4 and 7) the advantages of this way of coding will become even more apparent as we work with larger datasets held in relational databases and introduce approaches to *spatial* data science.

This chapter formally introduces the use of piping syntax with the dplyr package as a core framework for data and spatial data analyses. It extends the descriptions of tidy data and the steps involved in moving from data, to tidy data, to analysis and processing, through to communication of the results.

This chapter covers the following topics:

- Introduction to pipelines of tidy data
- The dplyr pipelining filters

- The tidy data chaining process

- Pipelines, `dplyr` and spatial data.

The following packages will be required for this chapter:

```
library(datasets)
library(tidyverse)
library(sf)
library(tmap)
```

Additionally, the IBDSDA package needs to be installed. This has been created by the authors of this book and has *not* been uploaded to the package repository hosted by the Comprehensive R Archive Network (CRAN) at https://cran.r-project.org. The zip file (ending in `.tar.gz`) should be downloaded from https://bit.ly/3axNdwS and then the package installed manually from the local zip file. Regardless of where the package is downloaded to, it will be compiled to the normal R library locations when it is installed. You may need to check or change your working directory for downloading the file. The current working directory can checked with:

```
getwd()
```

If it needs to be changed this can be done via **Session > Set Working Directory > Choose Directory**. If you have done these then the IBDSDA package can be loaded using the code below, **or** it can be done manually from the package archive (`.tar.gz`) file using the RStudio Menu: **Tools > Install Packages > Install from (dropdown) > Package Archive File**.

```
# download the package zip file
download.file("http://archive.researchdata.leeds.ac.uk/733/1/
              IBDSDA_0.1.2.tar.gz",
              "./IBDSDA_0.1.2.tar.gz", mode = "wb")
# install the package
install.packages("IBDSDA_0.1.2.tar.gz", type="source", repos=NULL)
```

Having done this the package can be loaded:

```
library(IBDSDA)
```

Finally, you will also need to re-create the `state_tidy` and `AirPassengers_tidy` data from Chapter 2:

```
# state_tidy
state_tidy <- as.data.frame(state.x77)
```

```
state_tidy$Division <- as.character(state.division)
state_tidy$Region <- as.character(state.region)
state_tidy$Name <- rownames(state.x77)
# AirPassengers_tidy
AirPassengers_df <- matrix(datasets::AirPassengers,12,12,
                           byrow = TRUE)
rownames(AirPassengers_df) <- 1949:1960
colnames(AirPassengers_df) <-c("Jan","Feb","Mar","Apr","May",
                               "Jun","Jul","Aug","Sep","Oct",
                               "Nov","Dec")
AirPassengers_df <- data.frame(AirPassengers_df)
AirPassengers_df$Year <- rownames(AirPassengers_df)
AirPassengers_tidy <- gather(AirPassengers_df,
                             key = Month, value = Count, -Year)
as_tibble(AirPassengers_tidy)
## # A tibble: 144 x 3
##      Year  Month Count
##      <chr> <chr> <dbl>
##  1 1949   Jan     112
##  2 1950   Jan     115
##  3 1951   Jan     145
##  4 1952   Jan     171
##  5 1953   Jan     196
##  6 1954   Jan     204
##  7 1955   Jan     242
##  8 1956   Jan     284
##  9 1957   Jan     315
## 10 1958   Jan     340
## # ... with 134 more rows
```

Key Points

- Methods for installing packages saved as local archives (.tar and .zip files) were introduced.

- This chapter introduces the use of piping syntax with the dplyr package, providing a core framework for data and spatial data analyses.

3.2 INTRODUCTION TO PIPELINES OF TIDY DATA

In Chapter 2 the code snippet below was used to filter out records (places) from the state_tidy data table where average life expectancy falls below 70 years.

Other code snippets nested a sequence of dplyr verbs to summarise mean income to illiteracy ratios for the nine divisions in the USA:

```
arrange(
  summarise(
    group_by(
      mutate(state_tidy, Inc_Illit = Income/Illiteracy),
      Division),
    mean_II = mean(Inc_Illit)),
  mean_II)
```

Note that the expression above is quite hard to read, as there are several layers of parentheses, and it is hard to see immediately which parameters relate to the individual dplyr verbs for sorting, filtering, summarising, grouping and arranging. The dplyr package presents an alternative syntax that overcomes this, using the pipeline operator %>%. Effectively this allows a variable (usually a tidy data frame) to be placed in front of a function so that f(x) is replaced by x %>% f. So, rather than the function f *being applied to* x, x is *piped into* the function. If there are multiple arguments for f, we have f(x,y,z,...) replaced by x %>% f(y,z,...) – thus typically the first argument is placed to the left of the function.

The reason why this can be helpful is that it makes it possible to specify chained operations of functions on tidy data as a left-to-right narrative. The pipelined versions of the above operations are:

```
state_tidy %>% arrange(Region,Division,`Life Exp`) %>%
  filter(`Life Exp` < 70)
```

and:

```
state_tidy %>%
  mutate(Inc_Illit = Income/Illiteracy) %>%
  group_by(Division) %>%
  summarise(mean_II = mean(Inc_Illit))
```

From the piped code snippets above, it is clear that the input data frame is state_tidy in both cases, and in the first that an arrange operator is applied, and then a filter. Since this is not stored anywhere else, the result is printed out. Similarly, the second code snippet first applies a mutate operation and then groups the result using group_by, before calculating a group summary.

Inspecting piped operations

One key thing to note in the construction of pipelines of operations is that the action of each operation in the pipeline can be inspected.

(Continued)

If the input is a data.frame then the function head() can be inserted (and then removed) after each %>% in the pipeline. So, for example, the above code could be inspected as follows:

```
state_tidy %>%
  mutate(Inc_Illit = Income/Illiteracy) %>% head()
```

If the input is a tibble then the code could be run to just before the pipe operator as the print function for tibble formats prints out the first 10 lines:

```
# make a tibble
state_tidy %>% as_tibble() %>%
  mutate(Inc_Illit = Income/Illiteracy)
```

It is possible – as is generally the case in R – to store the output in a new variable. The 'classic' R way of doing this would be as follows:

```
state_low_lifex <-
  state_tidy %>% arrange(Region, Division, `Life Exp`) %>%
  filter(`Life Exp` < 70)
```

However, for some this sits uneasily with the left-to-right flow using the piping operators as the final operation (assignment) appears at the far left-hand side. An alternative R assignment operator exists, however, which maintains the left-to-right flow. The operator is ->:

```
state_tidy %>% arrange(Region, Division, `Life Exp`) %>%
  filter(`Life Exp` < 70) -> state_low_lifex
```

The tibble format was introduced in Chapter 2. Recall that tibbles are basically data frames with a few modifications to help them work better in a tidyverse framework and that a standard data frame can be converted to a tibble with the function as_tibble:

```
state_tbl <- as_tibble(state_tidy)
state_tbl
```

```
## # A tibble: 50 x 11
##      Population Income Illiteracy `Life Exp` Murder `HS Grad`
##          <dbl>  <dbl>      <dbl>      <dbl>  <dbl>     <dbl>
## 1         3615   3624        2.1       69.0   15.1      41.3
## 2          365   6315        1.5       69.3   11.3      66.7
## 3         2212   4530        1.8       70.6    7.8      58.1
## 4         2110   3378        1.9       70.7   10.1      39.9
```

```
## 5      21198    5114          1.1          71.7    10.3        62.6
## 6       2541    4884          0.7          72.1     6.8        63.9
## 7       3100    5348          1.1          72.5     3.1        56
## 8        579    4809          0.9          70.1     6.2        54.6
## 9       8277    4815          1.3          70.7    10.7        52.6
## 10      4931    4091          2            68.5    13.9        40.6
## # … with 40 more rows, and 5 more variables: Frost <dbl>,
## #    Area <dbl>, Division <chr>, Region <chr>, Name <chr>
```

Key Points

- The dplyr package supports a piping syntax.

- This allows sequences of dplyr operations to be chained together with the pipeline operator %>%.

- In these left-to-right sequences, the output of one operation automatically becomes the input of to the next.

- The final operation can be assignment to an R object, and this can be done using alternative R assignment operators:

  - using -> at the end of the piped operations, which maintains the left-to-right flow;

  - using <- before the first piped operation, in a more 'classic' R way.

- The final operation can also *pipe* the results to other actions such as ggplot() (see the end of Chapter 2).

3.3 THE dplyr PIPELINING FILTERS

As well as filter and arrange, a number of other tidy-to-tidy transformations exist. In this section some key ones will be outlined.

3.3.1 Using select for column subsets

While filter selects subsets of rows from a data frame, select selects subsets of columns. Returning to state_tbl, a basic selection operation works as follows:

```
state_tbl %>% select(Region, Division, Name, `Life Exp`)
```

This selects the columns Region, Division, Name and Life Exp from state_tbl. Note also that ordering of columns matters:

```
state_tbl %>% select(Name, Division, Region, `Life Exp`)
```

It is possible to specify ranges of columns using:

```
state_tbl %>% select(Population:Murder)
```

This selects the range of columns from Population to Murder in the order they occur in state_tbl. Also, using a minus sign in front of a variable causes it to be *omitted*:

```
state_tbl %>% select(-Population)
```

You can also use the minus sign on ranges – although parentheses should be used:

```
state_tbl %>% select(-(Population:Murder))
```

There are also operations allowing selections based on the text in the column name. These are starts_with, ends_with, contains and matches. For example:

```
state_tbl %>% select(starts_with("Life"))
```

These are useful when a group of column names have some common pattern – for example, Age 0 to 5, Age 6 to 10 and so on could be selected using starts_with("Age"). matches is more complex, as it uses regular expressions to specify matching column names. For example, suppose the data also had a column called Agenda – this would also be selected using starts_with("Age") – but using matches("Age [0-9]") would correctly select only the first set of columns. There is also a function called num_range which selects variables with a common stem and a numeric ending – for example, num_range("Type_",1:4) selects Type_1, Type_2, Type_3 and Type_4.

The helper function everything() selects all columns. This is sometimes useful for changing the order in which columns appear. In a tidy data frame column order does not matter in terms of information stored – a consequence of the requirement that every variable has a column – but in presentational terms reordering is useful, for example so that some variables of current interest appear leftmost (i.e. first) when the table is listed. The following ensures the Name and then the Life Exp columns come first:

```
state_tbl %>% select(Name, `Life Exp`,everything())
```

The reason why this works is that Name and then Life Exp are selected in order followed by *all* of the columns in current order. This will include the previous columns, but if a column name is duplicated, the column only appears once, in the first position it is listed. So effectively everything() here means the rest of the columns in order.

A quick note on `select`. Lots of packages have a function named `select`, including for example `Raster`, `dplyr` and `MASS`. This is the most common cause of a `select` operation not working! If you have multiple packages containing a function named `select`, then you need to tell R which one to use. The syntax for doing this uses a double colon as follows:

```
state_tbl %>% dplyr::select(Name, `Life Exp`,everything())
```

3.3.2 Using mutate to derive new variables and transform existing ones

`mutate` can be used to transform variables in a data frame, or derive new variables. For example, to compute the logarithm (to base 10) of the `Income` column in `state_tbl`, use:

```
state_tbl %>% mutate(LogIncome=log10(Income)) %>%
    select(Name, Income, LogIncome)
```

To actually *overwrite* the `Income` column, enter:

```
state_tbl %>% mutate(Income=log10(Income))
```

Often it is a good idea to avoid overwriting unless the results are easy to undo – otherwise it becomes very difficult to roll back if you make an error. Here, no long-term damage was done since although the `mutate` operation was carried out and a tibble object was created with a redefined `Income` column, the new tibble was not stored anywhere – and, most importantly, it was not overwritten onto `state_tbl`.

A `mutate` expression can involve any of the columns in a tibble. Here, a new column containing income per head of population is created:

```
state_tbl %>% mutate(`Per Capita`=Income/Population) %>%
    select(Name, Income, Population, `Per Capita`)
```

The pipelining idea can be utilised to advantage here. To create the *per capita* income column and then sort on this column, reorder the columns to put the state name and per capita income columns to the left, storing the final result into a new tibble called `state_per_cap`, enter:

```
state_tbl %>% mutate(`Per Capita`=Income/Population) %>%
    arrange(desc(`Per Capita`)) %>%
    select(Name, `Per Capita`, everything()) -> state_per_cap
state_per_cap
```

```
## # A tibble: 50 x 12
##     Name   `Per Capita`  Population Income  Illiteracy
##     <chr>      <dbl>        <dbl>    <dbl>     <dbl>
##  1 Alas…      17.3          365     6315       1.5
##  2 Wyom…      12.1          376     4566       0.6
##  3 Neva…       8.73         590     5149       0.5
##  4 Dela…       8.31         579     4809       0.9
##  5 Verm…       8.28         472     3907       0.6
##  6 Nort…       7.99         637     5087       0.8
##  7 Sout…       6.12         681     4167       0.5
##  8 Mont…       5.83         746     4347       0.6
##  9 Hawa…       5.72         868     4963       1.9
## 10 New …       5.27         812     4281       0.7
## # … with 40 more rows, and 7 more variables: `Life
## #   Exp` <dbl>, Murder <dbl>, `HS Grad` <dbl>, Frost <dbl>,
## #   Area <dbl>, Division <chr>, Region <chr>
```

In non-pipelined format the above calculation would be:

```
state_per_cap <-
  select(
   arrange(
     mutate(state_tbl, `Per Capita`=Income/Population),
     desc(`Per Capita`)),
   Name, `Per Capita`, everything())
```

This is of course a matter of opinion, but the authors feel that the dplyr version is more easily understood.

Expressions can also contain variables external to the tibble. For example, we added the Division column to the original data frame state_tidy based on the external variable state.division using state_tidy$Division <- as.character(state.division) with standard R. A dplyr version of this would be:

```
state_tidy %>% mutate(Division=as.character(state.division)) ->
  state_tidy
state_tidy
```

A related operation is transmute. This is similar to mutate, but while mutate keeps the existing columns in the result alongside the derived ones, transmute *only* returns the derived results. For example:

```
state_tbl %>% transmute(LogIncome=log10(Income),
                        LogPop=log10(Population))
```

To keep some of the columns from the original, but without transforming them, just state the names of these columns:

```
state_tbl %>%
transmute(Name,Region,LogIncome=log10(Income),
          LogPop=log10(Population))
```

```
## # A tibble: 50 x 4
##    Name          Region     LogIncome   LogPop
##    <chr>         <chr>          <dbl>    <dbl>
##  1 Alabama       South           3.56     3.56
##  2 Alaska        West            3.80     2.56
##  3 Arizona       West            3.66     3.34
##  4 Arkansas      South           3.53     3.32
##  5 California    West            3.71     4.33
##  6 Colorado      West            3.69     3.41
##  7 Connecticut   Northeast       3.73     3.49
##  8 Delaware      South           3.68     2.76
##  9 Florida       South           3.68     3.92
## 10 Georgia       South           3.61     3.69
## # ... with 40 more rows
```

3.3 group_by AND summarise: CHANGING THE UNIT OF OBSERVATION

The state_tbl variable has rows corresponding to US states, and the columns are variables (statistics) relating to each state. However, there are other geographical units in the data, as specified in the Division and Region columns. The spatial units are nested, so that the Division units are subdivisions of Region, and the states (indicated by Name) are nested within divisions. There are times when it may be useful to use higher-level groupings such as Division or Region as a unit of observation, and aggregate variables to these units. This is where group_by and summarise are helpful.

A key idea here is that while mutate carries out operations that work row-by-row on columns (so that, for example, log10(Income) creates a column whose individual elements are the log (base 10) of the corresponding element in the column Income), summarise works with functions returning a single number for the whole column – such as mean, median or sum. Thus, for example:

```
state_tbl %>% summarise(TotPop=sum(Population),TotInc=sum(Income))
```

```
## # A tibble: 1 x 2
##    TotPop  TotInc
##     <dbl>   <dbl>
## 1  212321  221790
```

In this case the output is a tibble, but because sum maps a column onto a single number there is only one row, with elements corresponding to the two summarised columns. Essentially, because the sums are for the whole of the USA, the unit of observation is now the USA as a country.

However, we may wish to obtain the sums of these quantities by region. This is where group_by is useful. Applying this function on its own has little noticeable effect:

```
state_tbl %>% group_by(Region)

## # A tibble: 50 x 11
## # Groups:    Region [4]
##      Population Income Illiteracy `Life Exp` Murder `HS Grad`
##           <dbl>  <dbl>      <dbl>      <dbl>  <dbl>     <dbl>
##  1        3615   3624        2.1       69.0   15.1      41.3
##  2         365   6315        1.5       69.3   11.3      66.7
##  3        2212   4530        1.8       70.6    7.8      58.1
##  4        2110   3378        1.9       70.7   10.1      39.9
##  5       21198   5114        1.1       71.7   10.3      62.6
##  6        2541   4884        0.7       72.1    6.8      63.9
##  7        3100   5348        1.1       72.5    3.1      56
##  8         579   4809        0.9       70.1    6.2      54.6
##  9        8277   4815        1.3       70.7   10.7      52.6
## 10        4931   4091        2         68.5   13.9      40.6
## # ... with 40 more rows, and 5 more variables: Frost <dbl>,
## #   Area <dbl>, Division <chr>, Region <chr>, Name <chr>
```

The only visible effect is that when the tibble is printed it records that there are groups by region. Internally the variable specified for grouping is stored as an attribute of the tibble. However, if this information is then passed on to summarise then a new effect occurs. Rather than applying the summarising expressions to a whole column, the column is split into groups on the basis of the grouping variable, and the function applied to each group in turn.

Again, pipelining is a useful tool here – group_by is applied to a tibble, and the grouped tibble is piped into a summarise operation:

```
state_tbl %>% group_by(Region) %>%
  summarise(TotPop=sum(Population),TotInc=sum(Income))

## # A tibble: 4 x 3
##    Region          TotPop   TotInc
##    <chr>            <dbl>    <dbl>
## 1 North Central     57636    55333
## 2 Northeast         49456    41132
## 3 South             67330    64191
## 4 West              37899    61134
```

The output is still a tibble but now the rows are the regions. As well as the summary variables TotPop and TotInc the grouping variable (Region) is also included as a column. Similarly, we could look at the division as a unit of analysis with a similar operation:

```
state_tbl %>% group_by(Division) %>%
   summarise(TotPop=sum(Population),TotInc=sum(Income))
```

Using pipelines, a new tibble with divisional income per capita, sorted on this quantity, can be created. Again, procedures like this benefit most from dplyr. Here we store the result in a new tibble called division_per_capita:

```
state_tbl %>% group_by(Division) %>%
   summarise(TotPop=sum(Population),TotInc=sum(Income)) %>%
   mutate(`Per Capita`=TotInc/TotPop) %>% arrange(desc(`Per
   Capita`)) %>%
   select(Division,`Per Capita`,TotInc,TotPop) ->
   division_per_capita
division_per_capita
## # A tibble: 9 x 4
##    Division              `Per Capita`    TotInc    TotPop
##    <chr>                     <dbl>        <dbl>     <dbl>
## 1 Mountain                   3.66         35218      9625
## 2 New England                2.18         26543     12187
## 3 West North Central         1.92         31988     16691
## 4 South Atlantic             1.06         34842     32946
## 5 East South Central         1.05         14255     13516
## 6 Pacific                    0.917        25916     28274
## 7 West South Central         0.723        15094     20868
## 8 East North Central         0.570        23345     40945
## 9 Middle Atlantic            0.391        14589     37269
```

It is also possible to group by more than one variable. For example, a new logical variable called Dense is introduced here based on the state population density – the population divided by the area. More rural states will have a lower density. Here Dense is an indicator as to whether the density exceeds 0.25. Using this and Region as grouping columns, the median life expectancy for each group is compared, where a group is a specific combination of each of the grouping variables. Note the use of n() as a summary function. This provides the number of observations in a group, here assigned to the column N in the output tibble:

```
state_tbl %>% mutate(Dense=Population/Area > 0.25) %>%
   group_by(Dense,Region) %>%
   summarise(`Life Exp`=median(`Life Exp`), N=n())
```

```
## # A tibble: 7 x 4
## # Groups: Dense [2]
##    Dense Region            `Life Exp`      N
##    <lgl> <chr>                  <dbl> <int>
## 1 FALSE North Central         72.5     11
## 2 FALSE Northeast             71.2      3
## 3 FALSE South                 69.8     14
## 4 FALSE West                  71.7     13
## 5 TRUE  North Central         70.8      1
## 6 TRUE  Northeast             71.4      6
## 7 TRUE  South                 70.1      2
```

Another thing to see here is that there are seven rows of data, but actually eight possible Region–Dense combinations. However, a row is only created if a particular combination of the grouping variables occurs in the input data. In this case, there is no dense state in the West region, while all the other combinations do occur, hence the discrepancy of one row.

3.3.4 group_by with other data frame operations

Although perhaps the most usual use of group_by is with summarise, it is also possible to use it in conjunction with mutate (and transmute) and filter. For mutate this is essentially because it is possible to use column-to-single-value functions (such as mean, sum and max) as part of an expression. Suppose we wish to standardise the life expectancy values against a US national mean figure:

```
state_tbl %>% transmute(Name, sle = 100*`Life Exp`/mean
                        (`Life Exp`))
## # A tibble: 50 x 2
##    Name          sle
##    <chr>       <dbl>
##  1 Alabama      97.4
##  2 Alaska       97.8
##  3 Arizona      99.5
##  4 Arkansas     99.7
##  5 California  101.
##  6 Colorado    102.
##  7 Connecticut 102.
##  8 Delaware     98.8
##  9 Florida      99.7
## 10 Georgia      96.7
## # … with 40 more rows
```

A `transmute` operation is still sensible here because the summary statistic is the denominator in a one-to-one row expression, and the overall result is still one-to-one. The numbers themselves are standardised so that very quickly one can tell whether a particular state has a life expectancy above or below the national average. For example, Alabama is a little below.

However, suppose we wished to standardise against regional rather than national averages. If a tibble has been grouped, then operations such as sum apply to the group that each row belongs to rather than the entire dataset. Thus, to standardise at the regional level, the following can be used:

```
state_tbl %>% group_by(Region) %>%
  transmute(Name, region_sle = 100*`Life Exp`/mean(`Life Exp`))
```

Now the values have changed – for example, although Alabama still has a standardised life expectancy below 100, this is less marked when considered at the regional level. Also, as with `summarise` the grouping variable is implicitly carried through to the output tibble, so that even if the `transmute` command did not specify this column, it still appears in the output tibble.

As another example, arguably the mean value used in the calculation above should be population weighted. The `weighted.mean` function could be used to achieve this:

```
state_tbl %>% group_by(Region) %>%
  transmute(Name,
  region_sle = 100*`Life Exp`/weighted.mean(`Life Exp`, Population))
```

Another kind of function whose behaviour is modified by group is a *window function*. These are functions that return a new column, but the relationship between the input and output columns is not on a one-to-one basis. An example of a function that operates on a one-to-one basis is `log10` – the *i*th item of the output column is simply the log to base 10 of the *i*th item of the input column. Thus `log10` is not a window function. However, the function `rank` is – this returns the rank of each observation in a column. The rank of an individual item in a column depends not only on that item, but also on all the other items. To see the rank of life expectancy, one could enter:

```
state_tbl %>% transmute(Name, `Rank LE`=rank(`Life Exp`))
```

Tied values result in fractional ranks, and a rank of 1 is the lowest value. A simple trick to reverse rank orders (i.e. to give the maximum value a rank of 1) is to change the sign of the ranking column:

```
state_tbl %>% transmute(Name, `Rank LE`=rank(-`Life Exp`))
```

Window functions are also affected by grouping – in this case they are applied to each group separately – so if the data were grouped by region then the rank of Alabama would be in relation to the states in the South region, and so on. Assuming no ties, we would expect four rank 1 observations in the entire tibble since it is grouped into the four regions. Here is an example:

```
state_tbl %>% group_by(Region) %>%
  transmute(Name, `Rank LE`=rank(-`Life Exp`))
```

Other useful window functions are scale (which returns z-scores) and ntile. The latter assigns an integer value to each item in the column based on ordering the values into *n* evenly sized buckets. Thus, if there were 100 observations in column x, ntile(x,4) assigns 1 to the lowest 25 values, 2 to the next 25 and so on. Again, these will all be applied on a groupwise basis if group_by has been used. Here ntile is used regionally:

```
state_tbl %>% group_by(Region) %>%
  transmute(Name, `Illiteracy group`=ntile(Illiteracy,3)) %>%
  arrange(desc(`Illiteracy group`))
```

The filter function also respects grouping in this way. So, for example, to select all of the states whose illiteracy rate is above the mean for their region, use:

```
state_tbl %>% group_by(Region) %>%
  filter(Illiteracy > mean(Illiteracy)) %>%
  select(Name,Illiteracy,everything())
## # A tibble: 23 x 11
## # Groups:   Region [4]
##      Name   Illiteracy Population Income `Life Exp` Murder
##      <chr>      <dbl>      <dbl>  <dbl>      <dbl>  <dbl>
##   1 Alab...      2.1       3615   3624       69.0   15.1
##   2 Alas...      1.5        365   6315       69.3   11.3
##   3 Ariz...      1.8       2212   4530       70.6    7.8
##   4 Arka...      1.9       2110   3378       70.7   10.1
##   5 Cali...      1.1      21198   5114       71.7   10.3
##   6 Conn...      1.1       3100   5348       72.5    3.1
##   7 Geor...      2         4931   4091       68.5   13.9
##   8 Hawa...      1.9        868   4963       73.6    6.2
##   9 Illi...      0.9      11197   5107       70.1   10.3
## 10 Loui...      2.8       3806   3545       68.8   13.2
## # ... with 13 more rows, and 5 more variables: `HS
## #    Grad` <dbl>, Frost <dbl>, Area <dbl>, Division <chr>,
## #    Region <chr>
```

The effect of grouping on the operations select and mutate is more subtle. It does not cause any functions to be evaluated on a group-by-group basis but the grouping still remains, and the group column is always in the output, even when it is not explicitly specified.

Note one potential trap here: arrange does not take notice of groupings – so that arranging a grouped tibble by a variable will order it in terms of the ranking of a given variable for the *whole* column without noting which group an observation belongs to. However, the output retains the grouping of the input – so piping this into summarise would still apply on a group-by-group basis. However, it is possible to achieve this by specifying the grouping variable in an arrange operation with nested sorting:

```
state_tbl %>% group_by(Region) %>%
    arrange(Region,Illiteracy) %>% select(Name,Illiteracy)
```

3.3.5 Order-dependent window functions

The window functions shown so far, such as rank and scale, are order independent, in that if you were to permute the order in which the rows appear the values of the transformed cells would be permutated identically. If the observation ranked seventh in a column were moved to a different row in the table, then the value 7 would be moved to that new row. However, other window functions do not have this characteristic. For example, cumsum is a window function, returning the cumulative sum of values of observations from the first row to each individual row. A permutation as described above would alter this quantity. While this does not stop this being a useful function in many contexts, care must be taken to know the ordering of the data. The authors' recommendation is to only use such a function after an arrange operation to remove any ambiguity about the order of the rows.

A list of order-dependent window functions includes the following:

- lag returns a column of observations from the previous row – the first row has the value NA.

- lead returns a column of observations from the next row – the last row has the value NA.

- cumsum returns the cumulative sum.

- cummin returns the cumulative minimum.

- cummax returns the cumulative maximum.

- cummin returns the cumulative mean.

- cumall returns the cumulative logical 'all' function.

- cumany returns the cumulative logical 'any' function.

The first two have some further arguments. n indicates the size of the lag or lead (e.g. if data are sorted by month, and all months in a study period are present, then using n=12 provides the observation from one year ago) with the first year of observations mapping to NA. Returning to the air passenger data, this can be converted to a `tibble` and then used to compute the average annual increase on the previous year for each month:

```
AirPassengers_tbl <- as_tibble(AirPassengers_tidy)
AirPassengers_tbl %>% arrange(Year,Month) %>%
   transmute(Change=Count - lag(Count,n=12),Month) %>%
   group_by(Month) %>% summarise(Change=mean
                                (Change, na.rm=TRUE))
```

Note that in the `transmute` operation, the `Month` is added as well as the lagged changes so that it is carried through for the grouping operation. Here it was necessary to perform the grouping *after* this step as it was required to perform the lagging on the *ungrouped* data. Also, `na.rm=TRUE` had to be added to the `mean` function, as the first 12 values of `Change` were NA.

Key Points

- The `dplyr` package has a number of operations for pipelining and extracting data from data tables.

- The output of piped operations is in the same format as the input (`tibble`, `data.frame`, etc.).

- The `select` function subsets columns by name, by named ranges and pattern matching such as `starts_with`, and the minus sign in front of named variable causes it to be *omitted* (e.g. `select(-Population)`).

- Many packages have a function named `select`, and the `dplyr` version can be enforced using the syntax `dplyr::select()`.

- The `mutate` and `transmute` functions can be used to both create new (column) variables and transform existing ones.

- While `mutate` keeps the existing columns in the result alongside the derived ones, `transmute` *only* returns the derived results.

- The `summarise` function summarises numerical variables in a data table.

- The `group_by` function in conjunction with `summarise` allows single or multiple group summaries to be calculated.

3.4 THE TIDY DATA CHAINING PROCESS

The previous sections outlined the main functions used to transform and arrange data in the tidy format. There is a further set of operations to allow some more analytical approaches, but for now these provide a useful set of tools – and certainly provide a means for exploring and selecting data prior to a more formal analysis. The advantage of this set of methods is that they are designed to chain together so that the output from one method can feed directly as input to the next. This is due to some common characteristics – all of the pipelining operators take the form

```
output <- method(input,method modifiers...)
```

where `input` and `output` are data frames or tibbles, and `modifiers` provide further information for the `method`. Another characteristic is that the modifiers can refer to the column names in a data frame directly – that is, `col1` can be used rather than `df$col1`. Both of these are conventions rather than necessities, but make coding easier as users can depend on the functions behaving in a certain way. Although the method template above was in 'classic R' (or base R) format, in pipeline form it would be

```
input %>% method(method modifiers...) -> output
```

where the advantages become clear, as then one can use:

```
input %>% method1(method1 modifiers...) %>%
    method2(method modifiers2...) ...
```

However, this assumes that data are always supplied in tidy form, and that the final output required is a tidy tibble or data frame. Often this is not the case – data are provided in many forms, and although tidy data frames are easy to handle, the final output may well be more understandable and easily viewed in some other format. This might, for example, be a rectangular table with columns for months and rows for years (not tidy, but fits more neatly on a printed page or web browser window), or the output may require graphical representation via a graph or a map.

A more complete workflow than that proposed in Chapter 2 might be as follows:

1. Read in untidy data.

2. Make it tidy.

3. Perform a chain of tidy-to-tidy transforms with `mutate`, `filter`, etc.

4. Take tidy data and transform to an output to communicate results – typically a table or a graphic.

Good practice here is to separate the process into these distinct sets of operations in the given order. The data selection, transformation and possibly analysis should take place in step 3 (as a series of chained operations), whereas step 1 should be regarded as an extended 'reading-in' process. Step 4 should be regarded as transforming the final outcome of data processing in order to better communicate.

The whole process can then be thought of as an extended pipeline with an optional output as in Figure 3.1.

Each box or step in the pipeline may consist of multiple instructions, and certainly the middle box may be expressed explicitly in dplyr pipelined operations. A further characteristic is that the output of step 1, the input and output of step 2 and the input of step 3 should all be in tidy data format. Section 3.3 outlined a set of operators that may be used in step 2; in this section we focus on some tools for use in steps 1 and 3.

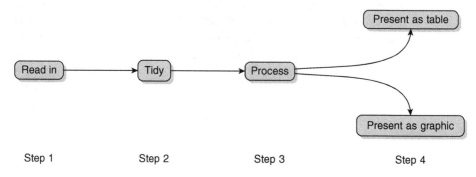

Figure 3.1   The recommended pipeline form for tidy data manipulation

3.4.1 Obtaining data

On many occasions, obtaining and reading in data is relatively straightforward – common file formats are .RData or .Rda, .csv, and .xls or .xlsx (MS Excel).

The first of these, .RData or .Rda, can be dealt with in base R via the load function. For the .csv case, base R provides the read.csv function. There is also a tidyverse version called read_csv, which in general is faster with large .csv files, automatically reads data into tibble format, leaves column names unchanged, and does not automatically convert character variables into factors by default. These were introduced in Chapter 2 but, in a nutshell, read_csv works on the principles of 'don't mess with the data any more than necessary' and 'read the data quickly'. Note that by default, read_csv guesses the data type of each column, and reports this when it reads in the data. The spec() function does the same thing and can be used to check the column data specification. The code below reloads the Sacramento real estate data introduced in Chapter 2:

```
real_estate_url="https://raw.githubusercontent.com/lexcomber/
                    bookdata/main/sac.csv"
real_estate = read_csv(real_estate_url)
```

You could confirm the specification:

```
spec(real_estate)
```

Suppose, however, it was thought that the zip column should be in character format. read_csv guessed it to be a double, because all zip codes are in fact valid numeric. However, in reality they behave more like characters (or factors) than numbers – for example, it makes no sense to add two zip codes together. This can be achieved in two ways. One is to use a basic R function called as.character which reads a non-character value and returns its character representation:

```
as.character(95838)
## [1] "95838"
```

Thus, we could use

```
real_estate %>% mutate(zip = as.character(zip)) -> real_estate2
real_estate2
```

to create a new tibble with zip as a character column. One issue here is that the spec characteristic of the tibble is now lost:

```
spec(real_estate2)
```

A better approach might be to specify the format of the columns during the read_csv operation, overriding the guesswork. This can be done with the col_types argument:

```
real_estate <- read_csv(real_estate_url,
                    col_types=cols(zip=col_character()))
```

The col_types argument essentially takes a list of column names and their types – the types specified by functions of the form col_integer(), col_character(), col_factor() and so on. Anything not specified in the list is still guessed. Some further columns where the guessing has been unhelpful are price and sq_ _ft which, although specified as <int> here (integers), may be more realistically thought of as <dbl> (double precision) – if, for example, they were to be used in regression modelling.

```
real_estate <- read_csv(real_estate_url,
                        col_types=cols(
                          zip=col_character(),
                          sq__ft=col_double(),
                          price=col_double()))
spec(real_estate)
## cols(
##    street = col_character(),
##    city = col_character(),
##    zip = col_character(),
##    state = col_character(),
##    beds = col_double(),
##    baths = col_double(),
##    sq__ft = col_double(),
##    type = col_character(),
##    sale_date = col_character(),
##    price = col_double(),
##    latitude = col_double(),
##    longitude = col_double()
## )
```

Once the data have been read in, a brief inspection using View(real_estate) raises some further issues. In particular, there are some properties with zero square feet floor area, zero bathrooms or zero bedrooms. These seem anomalous and so might be regarded as missing values. Similarly, there are a small number of properties with implausibly low prices (in particular, $1551, $2000 and several at $4897 all sold on the same date). This *could* be dealt with using filter or could be regarded as data tidying prior to processing. In the latter case, it could be handled by specifying that zero (or one of the rogue price values) is a missing value on numerical variables. There is another argument to read_csv called na which specifies which values are used for NA in the .csv file. Here, we could use:

```
real_estate <- read_csv(real_estate_url,
                        na=c("0","1551","2000","4897"),
                        col_types=cols(
                          zip=col_character(),
                          sq__ft=col_double(),
                          price=col_double()))
```

Another kind of file that is quite commonly encountered is the Excel file. MS Excel has been referred to as a tool for converting .xls and .xlsx files to .csv format. However, this approach is not reproducible as the procedures to do this are generally not scripted. A common issue is that spreadsheets are often format-ted in more complex ways than standard .csv files – several datasets on the same sheet, or a single .xlsx file with several worksheets each with a different dataset, are common. Also there is a tendency to merge data and metadata, so that notes on the definition of variables and so on appear on the same sheet as the data themselves.

The general consequence of this is that getting data out of Excel formatted files generally involves selecting a range of cells from a sheet, and writing these out as a .csv file. However, some reproducible alternatives exist, via R packages that can read .xls and .xlsx files. Here we introduce one of these – the package readxl. This offers a function called read_excel, taking the form read_excel(path, worksheet, range) in its basic form. path is simply the path to the Excel spreadsheet.

As an example, the Police Service of Northern Ireland (PSNI) releases crime data in spreadsheet form. This may be downloaded from the site https://www. psni.police.uk/inside-psni/Statistics (in particular, the data for November 2017 can be found in the file with URL https://www.psni.police.uk//globalassets/inside-the-psni/our-statistics/police-recorded-crime-statistics/2017/november/monthly-crime-summary-tables-period-ending-nov-17.xls). read_excel does not read directly from a URL, although it may be helpful to download the file and view it before extracting data in any case. This can be done via the download.file() operation. Note that the code below splits the URL into stem1 and stem2 for the purposes of the book!

```
stem1 <- "https://www.psni.police.uk//globalassets/inside-the-psni/"
stem2 <- "our-statistics/police-recorded-crime-statistics"
filepath <- "/2017/november/monthly-crime-summary-tables-period-
ending-nov-17.xls"
download.file(paste0(stem1,stem2,filepath),"psni.xls")
```

This downloads the file, making a local copy in your current folder called psni.xls. Opening this and looking at the sheet called Bulletin Table 2 shows a typical Excel table (Figure 3.2). As suggested earlier, this sheet contains footnotes and other metadata as well as the data themselves. In addition, it contains merged columns, sub-headings embedded in rows, and column names spread over multiple rows – a fairly typical mixture of presentation and data storage found in Excel files.

Here, the data seem to be in the range A5:E31 – although things will still be messy when the raw information is extracted. In any case, the first step is to read this in using read_excel():

```
library(readxl)
crime_tab <- read_excel(path="psni.xls",
                        sheet="Bulletin Table 2",
                        range="A5:E31")
print(crime_tab, n = 26)
```

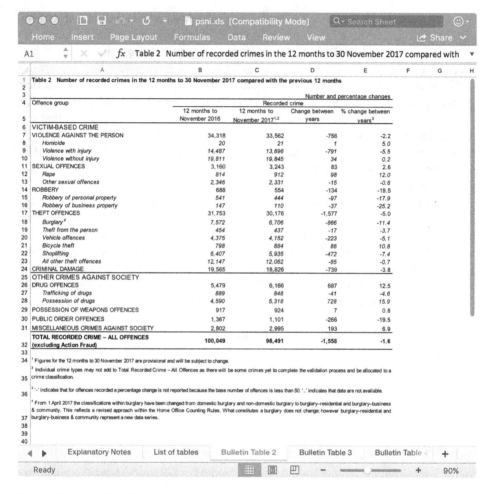

Figure 3.2 Part of an Excel spreadsheet

A number of issues can be seen here:

1. Some of the column names are unwieldy – and incorporate things like footnote references.

2. One column name was blank in the range A5:E31 (as it was on the row above).

3. Subtotals masquerade as ordinary rows (e.g. THEFT OFFENCES is the sum of the next six rows).

4. Some of the row names are really sub-headings (e.g. OTHER CRIMES AGAINST SOCIETY).

Both 3 and 4 above violate the tidy convention that each variable has its own column – basically there are nested categorisations of crime type – but rather than giving each level of categorisation its own column, the information is injected into a single column. Thus although the information is now in R, it is far from tidy. In the following subsection this issue will be considered.

3.4.2 Making the data tidy

Regardless of how the data got into R, quite frequently they are not in a tidy format. Recall from Chapter 2 that the requirements for tidy data are as follows:

1. Each variable must have its own column.

2. Each observation must have its own row.

3. Each value must have its own cell.

Returning to the *untidy* air passenger data, recall that the variable Month was spread out over 12 columns. When this data frame was introduced (as Air-Passengers_df) it was tidied without explanation. However, it made use of a tidyverse tool when doing this, the function gather(). This 'pulls in' multi-column variables, stacking the values in each input column and returning a single output column.

```
AirPassengers_df %>% gather(key="Month",
                            value="Count",-Year) %>% head
##      Year Month  Count
## 1 1949    Jan    112
## 2 1950    Jan    115
## 3 1951    Jan    145
## 4 1952    Jan    171
## 5 1953    Jan    196
## 6 1954    Jan    204
```

Here, the key argument provides the name given to the variable derived from the multiple column names when the column stacking occurs (essentially it just indicates which of the columns the newly stacked data were taken from). The value argument provides the name for the column of stacked values – here Count, as it is a passenger count. Finally, the last argument specifies the column range over which the gathering takes place. Here we want to gather everything except Year. Columns can be specified in the same way as select so -Year specifies gathering over all of the columns except Year. Any non-gathering columns will simply have their values stacked in correspondence to their row in the staking range – so the row 1 value (1949) will appear next to every gathered value from that row, and so on.

The data are now tidy, although before processing them, we may wish to transform some of the data types. This was done in base R initially, but now it may be done using `mutate` with `dplyr` functions to convert between types including `parse_character`, `parse_integer` and `parse_factor`. Suppose here we would like Year and Count to be integers (at the moment Year is character and Count is double precision), and Month to be a factor (so that it tabulates in month order, not alphabetically). It might also be useful to convert the base R data frame into a tibble. Using pipelines, the gathering and transformation may be carried out as follows:

```
AirPassengers_df %>% gather(key="Month",value="Count",-Year) %>%
  as_tibble %>%
  mutate(Year=parse_integer(Year),
         Month=parse_factor(Month,
             levels=c("Jan","Feb","Mar","Apr","May","Jun",
                     "Jul","Aug","Sep","Oct","Nov","Dec")),
         Count=as.integer(Count))
```

For all of the `parse_<something>` variables except characters, it is possible that the input cannot be successfully parsed, for example `parse_integer("Maynooth")` makes no sense. In this case a value of NA is returned. This is reported when the supplied code has this issue – try entering the following:

```
flawed_data <-
  AirPassengers_df %>% gather(key="Month",value="Count",-Year) %>%
  as_tibble %>%
  mutate(Year=parse_integer(Year),
         Month=parse_integer(Month),
         Count=as.integer(Count))
```

The correct code (with the appropriate `parse_<something>` functions) has created a tidy data frame, with the column data types as required – effectively completing step 2 in Figure 3.1. Since this is in tidy format, some step-3-type analysis could also be carried out here. As a simple example, we could compute yearly total passenger counts:

```
AirPassengers_df %>% gather(key="Month",value="Count",-Year) %>%
  as_tibble %>%
  mutate(Year=parse_integer(Year),
         Month=parse_factor(Month,
             levels=c("Jan","Feb","Mar","Apr","May","Jun",
                     "Jul","Aug","Sep","Oct","Nov","Dec")),
         Count=as.integer(Count)) %>%
  group_by(Year) %>% summarise(Count=sum(Count))
```

As another example, using grouping and window functions, the passenger counts as percentages of the year in which they occur can also be computed:

```
AirPassengers_df %>% gather(key="Month",value="Count",-Year) %>%
   as_tibble %>%
   mutate(Year=parse_integer(Year),
          Month=parse_factor(Month,
                 levels=c("Jan","Feb","Mar","Apr","May","Jun",
                          "Jul","Aug","Sep","Oct","Nov","Dec")),
          Count=as.integer(Count)) %>%
   group_by(Year) %>%
   mutate(`Pct of Yearly Total`=100*Count/sum(Count))
```

In both cases the convention that step 2 (in Figure 3.1) outputs tidy data, and step 3 expects it as input, means that the whole process can be chained as a single pipeline.

Revisiting the Northern Ireland crime data, the situation is more complex, as the data are messier, with a number of issues. The first two of these concerned the column names. Here, transmute can be used as a tool to rename the columns – simply by defining new columns (with better names) to equate to the existing columns. First, check out the existing names:

```
colnames(crime_tab)
## [1] "...1"
## [2] "12 months to November 2016"
## [3] "12 months to \nNovember 20171,2"
## [4] "Change between years"
## [5] "% change between years3"
```

The ...1 name was provided since the column name for this column was in a different row on a spreadsheet than the others. Using transmute, we have:

```
crime_tab <- crime_tab %>%
   transmute(`Crime Type`=`...1`,
             `Year to Nov 2016`=`12 months to November 2016`,
             `Year to Nov 2017`=`12 months to \nNovember 20171,2`,
             `Change`=`Change between years`,
             `Change (Pct)`=`% change between years3`)
crime_tab
```

This addresses problems 1 and 2 in the list.

The next two problems relate to the fact that the unit of observation is Crime Type, but these are grouped into two broad classes (VICTIM-BASED CRIME and OTHER CRIMES AGAINST SOCIETY) and divided into several subclasses. For example, two subclasses of VICTIM-BASED CRIME are VIOLENCE AGAINST THE PERSON and SEXUAL OFFENCES. At the moment these appear as individual units of observation, so that several variables share the same column – with the broad classes having NA across all columns in their allocated rows. First, it would be useful to create a new column identifying which broad class each row belongs to. As there are just two classes, and everything after the row following the one with Crime Type as OTHER CRIMES AGAINST SOCIETY is in that broad class, and everything prior to it is in VICTIM-BASED CRIME, the window function cumany() is useful. The expression cumany(``Crime Type``=="OTHER CRIMES AGAINST SOCIETY") is TRUE for all observations following the critical row. Thus, an extra column can be added making use of this:

```
crime_tab %>%
  mutate(`Broad class` = ifelse
    (cumany(`Crime Type`=="OTHER CRIMES AGAINST SOCIETY"),
      "Other",
      "Victim-Based"))
```

```
## # A tibble: 26 x 6
##      `Crime Type`    `Year to Nov 20… `Year to Nov 20… Change
##      <chr>                    <dbl>            <dbl>  <dbl>
##   1 VICTIM-BASE…               NA               NA     NA
##   2 VIOLENCE AG…            34318            33562   -756
##   3 Homicide                   20               21      1
##   4 Violence wi…            14487            13696   -791
##   5 Violence wi…            19811            19845     34
##   6 SEXUAL OFFE…             3160             3243     83
##   7 Rape                      814              912     98
##   8 Other sexua…             2346             2331    -15
##   9 ROBBERY                   688              554   -134
## 10 Robbery of …              541              444    -97
## # … with 16 more rows, and 2 more variables: `Change
## #   (Pct)` <dbl>, `Broad class` <chr>
```

To complete this part of the tidying, the next step is to remove the rows corresponding to the broad classes. Noting that these are the only ones to contain NA for any cell, it is possible to pick any column, and filter on that, not taking NA as a value. Adding this filter to the pipeline, we have:

```
crime_tab %>%
  mutate(`Broad class` = ifelse(
    cumany(`Crime Type`=="OTHER CRIMES AGAINST SOCIETY"),
      "Other",
      "Victim based")) %>%
  filter(! is.na(Change))
```

```
## # A tibble: 24 x 6
##     `Crime Type` `Year to Nov 20… `Year to Nov 20… Change
##     <chr>                   <dbl>            <dbl> <dbl>
##   1 VIOLENCE AG…            34318            33562  -756
##   2 Homicide                   20               21     1
##   3 Violence wi…            14487            13696  -791
##   4 Violence wi…            19811            19845    34
##   5 SEXUAL OFFE…             3160             3243    83
##   6 Rape                      814              912    98
##   7 Other sexua…             2346             2331   -15
##   8 ROBBERY                   688              554  -134
##   9 Robbery of …              541              444   -97
## 10 Robbery of …              147              110   -37
## # … with 14 more rows, and 2 more variables: `Change
## #   (Pct)` <dbl>, `Broad class` <chr>
```

Next, the subgroups need addressing. One thing that distinguishes these in the CRIME TYPE column is that they are all in upper case. A test for this tells us that the row corresponds to a subgroup, not an individual crime type. The function toupper() converts a character string to all upper case:

```
toupper("Violence with injury")

## [1] "VIOLENCE WITH INJURY"
```

A logical expression of the form x == toupper(x) therefore tests whether x is already entirely upper case. Thus, the expression

```
`Crime Type` == toupper(`Crime Type`)
```

in a mutate operation adds a logical column that has the value TRUE if it corresponds to a subclass and FALSE otherwise. A neat trick here is to apply the cumsum() window function. This treats logical values as either 0 (for FALSE) or 1 (for TRUE). Thus it increases by 1 every time a row is a subclass and each row in the column has an integer from 1 to 9 indicating the subclass. The actual names (slightly rephrased and not in upper case) are stored in a vector called subclass_name which is indexed by the above calculation, providing a column of subclass names.

However, a further complication is that there are subclasses that have only one component – themselves. An example is PUBLIC ORDER OFFENCES. One way of dealing with these is to identify them using the add_count() operator. This is a one-off function that, given a column name, provides a new column n that counts the number of times each value in the column appears in the column as a whole. Thus if we have created a Subclass column, add_count(Subclass) (in pipeline mode) adds a new column n that counts the number of times the subclass value in the row appears in the column as a whole. If there are crime types in a given subclass, then every row with that subclass value will have an n value of $k + 1$ for the k members of the group, plus the row allocated to the subclass. The exceptions to this are the one-component subclasses, which only appear as a subclass row. In this case $k = 1$.

Finally, the *rows* referring to subclasses with more than one crime type are filtered out using the `toupper()` comparison method used before, and checking whether the n column is equal to 1. Having done this, n is no longer a useful column, and so it is removed via `select`:

```
subclass_name = c("Violence against the person",
                  "Sexual Offence","Robbery", "Theft",
                  "Criminal Damage","Drugs",
                  "Possession of Weapons", "Public Order",
                  "Misc")
crime_tab %>%
  mutate(`Broad class` = ifelse(
    cumany(`Crime Type`=="OTHER CRIMES AGAINST SOCIETY"),
    "Other",
    "Victim based")) %>%
  filter(! is.na(Change)) %>%
  mutate(Subclass = subclass_name[cumsum(
      `Crime Type` == toupper(`Crime Type`))]) %>%
  filter(`Crime Type` != toupper(`Crime Type`))
```

A final cleaning operation is required where a footnote in the Crime Type row on the spreadsheet causes "Burglary" to appear as "Burglary4". A final pipelining task can correct this with a 'brute force' substitution.

This is all quite complicated – it may be helpful to outline the necessary operations as a list:

1. Identify the broad class crime type and create a new column.

2. Remove the *rows* referring to broad class.

3. Identify the subclass crime type and create a new column.

4. Use `add_count` to see how many crime types are in a subclass.

5. Remove the *rows* corresponding to subclasses with more than one crime type.

6. Remove the n column.

7. Remove the footnote number from Burglaries.

This may then be carried out using the pipeline below:

```r
subclass_name = c("Violence against the person",
                  "Sexual Offence","Robbery", "Theft",
                  "Criminal Damage", "Drugs",
                  "Possession of Weapons", "Public Order",
                  "Misc")
crime_tab %>%
  mutate(`Broad class` = ifelse(
    cumany(`Crime Type`=="OTHER CRIMES AGAINST SOCIETY"),
    "Other",
    "Victim based")) %>%
  filter(! is.na(Change)) %>%
  mutate(Subclass = subclass_name[cumsum(
      `Crime Type` == toupper(`Crime Type`))]) %>%
  add_count(Subclass) %>%
  filter(`Crime Type` != toupper(`Crime Type`) | n == 1) %>%
  select(-n) %>%
  mutate(`Crime Type` = sub("4","", `Crime Type`)) -> crime_tidy
crime_tidy
## # A tibble: 19 x 7
##      `Crime Type` `Year to Nov 20… `Year to Nov 20… Change
##      <chr>                   <dbl>            <dbl>  <dbl>
##  1 Homicide                    20               21      1
##  2 Violence wi…             14487            13696   -791
##  3 Violence wi…             19811            19845     34
##  4 Rape                       814              912     98
##  5 Other sexua…              2346             2331    -15
##  6 Robbery of …               541              444    -97
##  7 Robbery of …               147              110    -37
##  8 Burglary                  7572             6706   -866
##  9 Theft from …               454              437    -17
## 10 Vehicle off…              4375             4152   -223
## 11 Bicycle the…               798              884     86
## 12 Shoplifting               6407             5935   -472
## 13 All other t…             12147            12062    -85
## 14 CRIMINAL DA…             19565            18826   -739
## 15 Trafficking…               889              848    -41
## 16 Possession …              4590             5318    728
## 17 POSSESSION …               917              924      7
## 18 PUBLIC ORDE…              1367             1101   -266
## 19 MISCELLANEO…              2802             2995    193
## # … with 3 more variables: `Change (Pct)` <dbl>, `Broad
## #   class` <chr>, Subclass <chr>
```

This demonstrates how the pipelining approach can be helpful in tidying very messy data. It also demonstrates why at least both of the authors' hearts sink when they are told they will be receiving some data as an Excel file. However, if you are compelled to use spreadsheets as a means of data sharing, Broman and Woo (2017) is worth reading.

To demonstrate again how this may flow into step 3 in Figure 3.1, the 2016 count for each crime time is computed as a percentage of the count for its subclass below:

```
crime_tidy %>% group_by(Subclass) %>%
    transmute(`Crime Type`,
              `Pct of Subclass`=
                 100*`Year to Nov 2016`/sum(`Year to Nov 2016`))
```

At this point it is perhaps worth reinforcing that getting the data into a format we wish to work with requires quite a lot of data 'munging' – reformatting and providing some information not supplied perfectly in the first place. This perhaps is a characteristic that should be highlighted in data science and that has been underplayed in traditional statistics: dealing with imperfect data can account for a great deal of time in practice, and although there are notable exceptions in the literature such as Chatfield (1995), much discussion of new analytical techniques is traditionally presented with data that magically arrive in an ideal form.

Key Points

- The pipeline chaining process and workflows require data to be *tidy*, but data as supplied may often not have tidy characteristics (see also Chapter 2).

- A number of issues related to different data formats were illustrated, including an untidy Excel example.

- A formal set of requirements for tidy data was outlined and applied: each variable must have its own column, each observation must have its own row, and each value must have its own cell.

- The use of the gather() function to do this was illustrated. This 'pulls in' multi-column variables, stacking the values in each input column and returning a single output column.

- Excel data were used to illustrate the need for extensive reformatting, reinforcing the key point that is often overlooked in data science that dealing with imperfect data can take a lot of time and effort.

3.5 PIPELINES, dplyr AND SPATIAL DATA

The sections above outline how dplyr and the concept of tidy data provide a framework in general for data manipulation. Chapter 2 provided some cursory examples of how dplyr functions could be applied to spatial data, and these are expanded here. Recall that the sp format for spatial data does not fit directly into the tidy framework, but the sf (simple features) framework does, as described in Chapter 2.

Here we will make use of the Newhaven burglary data supplied with the IBDSDA package. This contains several sf objects, including:

- blocks – US Census blocks area

- tracts – US Census tracts area

- crimes – report crimes: family disputes, breaches of the peace, residential burglary (forced/unforced)

- roads – roads in the area.

These relate to a number of crimes recorded in New Haven, CT, and to census data for the same area.

Each of these objects is actually a named item in a list object, called newhaven. First, these are read in:

```
library(IBDSDA)
data(newhaven)
```

Rather than having to refer to each of these items by their full name (e.g. newhaven$blocks) the list items are first copied out to separate individual objects:

```
tracts <- newhaven$tracts
roads <- newhaven$roads
callouts <- newhaven$callouts
blocks <- newhaven$blocks
```

You should examine these in the usual way. The tracts and blocks objects are census areas at different scales with attributes from the US Census. The roads layer contains different classes of road linear features, and the callouts data contain information on the locations of police callouts to different types of incidents.

These objects can be mapped using tmap, and the code below generates Figure 3.3:

```
library(tmap)
tmap_mode('plot')
tm_shape(blocks) + tm_borders() +
  tm_shape(roads) + tm_lines(col='darkred') +
  tm_shape(callouts) + tm_dots(col="Callout",size=0.5,alpha=0.3) +
  tm_scale_bar(position=c('left','bottom')) + tm_compass() +
  tm_layout(legend.position = c('left',0.1))
```

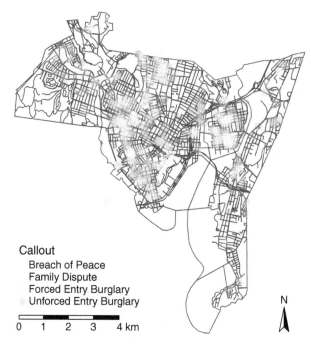

Figure 3.3 The *newhaven sf* objects mapped using *tmap*

3.5.1 dplyr and sf format spatial objects

Chapter 2 illustrated how sf objects can be manipulated in the dplyr framework. The code below filters out highways from the roads sf data. It has a column called AV_LEGEND specifying the type of road for each item in the object. These may be listed:

```
unique(roads$AV_LEGEND)
```

The type HWAY PRIM represents a primary highway. The filter() function can be used to select these roads:

```
roads %>% filter(AV_LEGEND=='HWAY PRIM') -> p_highway
p_highway
```

The different dplyr functions for manipulating data tables can be used to manipulate spatial data tables such as are held in sf objects. The code below calculates a count of rented but occupied properties, groups the data by tracts (blocks are nested within tracts in the US Census), summarises vacant and rented but occupied properties over the groups and creates a ratio, before passing the result to tmap:

```
blocks %>%
  mutate(RentOcc = HSE_UNITS*P_RENTROCC/100 ) %>%
  group_by(TRACTBNA) %>%
  summarise(RentOcc=sum(RentOcc),TotVac=sum(VACANT)) %>%
  mutate(Vac2Rent=TotVac/RentOcc) %>%
  tm_shape() +
  tm_fill(col = c("RentOcc", "TotVac", "Vac2Rent"),
          style = "kmeans",
          title =c("Rent Occ.", "Vacant Props.", "Ratio"),
          palette = "YlGnBu", format = "Europe_wide") +
  tm_layout(legend.position = c("left", "bottom"))
```

3.5.2 A practical example of spatial data analysis

It is now possible to formulate a geographical hypothesis around the linkage between the mode of entry of a burglary and location. Initially we will use a 'broad brush' definition of location – based on the census tracts. To do this, the following approach is suggested:

1. Identify which tract each burglary occurs in.

2. Count the number of forced/not forced burglaries in each tract.

3. Use a binomial generalised linear model to test the hypothesis that the probability of a residential burglary being via forced entry is the same in each tract.

The first two steps work very naturally in the dplyr tidy data framework. A useful function which can be thought of as a spatial addition to the left_join() function is the st_join() function in the sf package. Here, a simple features object is joined to another via a spatial linkage. For example, if the first argument consists of point geometry, and the second consists of polygons, the result is a new point-based object, with the attributes of the polygon containing each particular point being joined as columns.

Thus, the following code joins the `tracts` (polygon-based) data to the `callouts` (point-based) object:

```
callouts %>% st_join(tracts) %>% colnames
```

Note here that the original `callouts` object's columns are merged using the `tract` column (TRACTBNA). This provides a 'block area number' – a way of identifying tracts and blocks (now not used officially). Here it serves the purpose of identifying which tract each incident occurred in.

Suppose now we wish to focus on burglaries. This can be achieved by filtering via the `Callout` column. The four classes of callout include two relating to burglary, and their levels contain the word 'burglary'. The `str_detect()` function in the `stringr` package (which is automatically loaded when `tidyverse` is loaded) detects the presence of a character pattern in a string. If it is found, the function value is TRUE, otherwise FALSE. Thus it may be used as a filtering function to pick out the callouts relating to burglary. Next, `fct_drop` in the `forcats` package is used to redefine the factor `Callout` so it only has the levels that actually appear (i.e. only the levels relating to burglary). This is important to stop future tabulation and counting functions considering other levels, which have now been systematically excluded – at least for this particular analysis. The following code applies the filtering process outlined above to create an object called `burgres`:

```
callouts %>%
  filter(str_detect(Callout,"Burglary")) %>%
  mutate(Callout = fct_drop(Callout)) ->
  burgres
```

When `dplyr`-type functions are applied to `sf` objects, the `geometry` column is 'sticky' – that is, it does not disappear in operations where it would logically be expected to (such as after `transmute`, `select` or `summarise` operations). This is demonstrated below – even when `select` is used to drop the `geometry` column, it does not go away:

```
burgres %>% select(-geometry) %>% colnames
## [1] "Callout" "geometry"
```

To overcome this and get rid of this column, the `sf` object must first be converted to an ordinary tibble via the `as_tibble()` function. Thus to create a table with *just* the `Callout` and TRACTBNA columns, use:

```
burgres %>%
  st_join(tracts) %>%
  as_tibble %>%
  select(-geometry) ->burgres_tab
```

```
burgres_tab
```

```
## # A tibble: 220 x 2
##     Callout                   TRACTBNA
##     <fct>                     <fct>
##  1 Forced Entry Burglary 1403
##  2 Forced Entry Burglary 1427
##  3 Forced Entry Burglary 1407
##  4 Forced Entry Burglary 1403
##  5 Forced Entry Burglary 1421
##  6 Forced Entry Burglary 1424
##  7 Forced Entry Burglary <NA>
##  8 Forced Entry Burglary 1409
##  9 Forced Entry Burglary 1412
## 10 Forced Entry Burglary 1409
## # … with 210 more rows
```

Alternatively, the sf function st_drop_geometry can be used with the same effect. The sf package contains many 'pipeable' functions:

```
burgres %>%
  st_join(tracts) %>%
  st_drop_geometry() -> burgres_tab
```

The burgres_tab is not a spatial (sf) object, and cannot be used in spatial operations. However, it provides useful spatially derived data – here the count() function in dplyr is used to count the number of each type of burglary in each tract (ignore the warning!).

```
burgres_tab %>% count(Callout, TRACTBNA)
```

```
## # A tibble: 54 x 3
##     Callout                   TRACTBNA       n
##     <fct>                     <fct>      <int>
##  1 Forced Entry Burglary 1402          2
##  2 Forced Entry Burglary 1403          3
##  3 Forced Entry Burglary 1404          6
##  4 Forced Entry Burglary 1405          7
##  5 Forced Entry Burglary 1406          6
##  6 Forced Entry Burglary 1407         10
##  7 Forced Entry Burglary 1408         11
##  8 Forced Entry Burglary 1409         11
##  9 Forced Entry Burglary 1410          2
## 10 Forced Entry Burglary 1411          3
## # … with 44 more rows
```

The count() function is basically shorthand for

```
burgres_tab %>%
   group_by(Callout,TRACTBNA) %>%
   summarise(n=n()) %>%
   ungroup
```

where although the columns passed to group_by may be varied, the n column name in the result is fixed. It can be renamed by transmute or rename if an alternative name is preferred. However, an issue here is that Callout–TRACTBNA combinations that do not appear in the data do not appear as a grouping for summarising – this is always the case with grouping in dplyr. To overcome this, the complete() function is useful. Given a tibble or data frame with a list of columns in which only some combinations appear, it returns a new table with the missing combinations added as extra rows. The fill argument provides a list of the column names to be filled (when new rows are added), and the value to be filled. Thus, here, extra combinations of TRACTBNA and Callout not appearing are added as columns, but with n set to zero. This is important in the Poisson regression – otherwise zero count observations are excluded from the model, and information about unusual Callout–TRACTBNA combinations is lost.

```
burgres %>%
   st_join(tracts) %>%
   st_drop_geometry() %>%
   count(TRACTBNA,Callout) %>%
   complete(TRACTBNA,Callout,fill=list(n=0L)) ->
   burgres_counts
burgres_counts

## # A tibble: 62 x 3
##     TRACTBNA    Callout                      n
##     <fct>       <fct>                    <int>
##  1  1401        Forced Entry Burglary        0
##  2  1401        Unforced Entry Burglary      2
##  3  1402        Forced Entry Burglary        2
##  4  1402        Unforced Entry Burglary      1
##  5  1403        Forced Entry Burglary        3
##  6  1403        Unforced Entry Burglary      0
##  7  1404        Forced Entry Burglary        6
##  8  1404        Unforced Entry Burglary      2
##  9  1405        Forced Entry Burglary        7
## 10  1405        Unforced Entry Burglary      4
## # ... with 52 more rows
```

The fill value in complete is given as 0L rather than just 0 – the L after a number tells R that the number should be stored as a *long integer* (the default would be as a *double precision* real number), but since values here are counts, an integer format seems more appropriate.

It is now possible to use the output from this as data for a Poisson regression model. The two factors for predicting the count are Callout and TRACTBNA. The use of statistical modelling is covered in detail later in this book, but for now we will just state that the model is

$$\log(E(n)) = \alpha + \beta_i + \gamma_j + \delta_{ij} \tag{3.1}$$

where $E(\alpha)$ is the expected value of n, β_i is the effect due to crime type i, γ_j is the effect due to tract j, and δ_{ij} is the effect due to interaction between crime type i and tract j. If there is no interaction, this suggests that there is no difference (on average) between the rate of each type of burglary across the tracts, and so no evidence for geographical variability. In this case the simpler model

$$\log(E(n)) = \alpha + \beta_i + \gamma_j \tag{3.2}$$

is more appropriate. Here an analysis of deviance for the two models in equations (3.1) and (3.2) is carried out:

```
full_mdl <-
   glm(n~TRACTBNA*Callout,data=burgres_counts,family = poisson())
main_mdl <-
   glm(n~TRACTBNA+Callout,data=burgres_counts,family = poisson())
anova(main_mdl,full_mdl)

## Analysis of Deviance Table
##
## Model 1: n ~ TRACTBNA + Callout
## Model 2: n ~ TRACTBNA * Callout
##    Resid. Df Resid. Dev Df Deviance
## 1         29     37.586
## 2          0      0.000 29   37.586
```

Although there are reasons why significance testing should be treated with caution, as a guideline here note that the deviance of the full model (with δ_{ij}) is asymptotically χ^2 with Df (= 29) degrees of freedom. The upper one-tailed 5% threshold for this is 42.6, suggesting no evidence to reject a null hypothesis of independence here.

It is also possible to carry out a similar analysis for the other callout types – family disputes and breaches of the peace. Arguably these could both be termed 'disturbances'. Here the selection of the subset of callouts is the logical negation of the previous example. Other than that, all of the stages carried out before can be merged into a single dplyr pipeline, creating a new table called disturb_counts:

```
callouts %>%
   filter(!str_detect(Callout,"Burglary")) %>%
   mutate(Callout = fct_drop(Callout)) %>%
   st_join(tracts) %>%
   st_drop_geometry() %>%
   count(TRACTBNA,Callout) %>%
   complete(TRACTBNA,Callout,fill=list(n=0L)) ->
   disturb_counts
```

The Poisson regression can be applied in a similar way:

```
full_mdl2 <-
   glm(n~TRACTBNA*Callout,data=disturb_counts,family = poisson())
main_mdl2 <-
   glm(n~TRACTBNA+Callout,data=disturb_counts,family = poisson())
anova(main_mdl2,full_mdl2)

## Analysis of Deviance Table
##
## Model 1: n ~ TRACTBNA + Callout
## Model 2: n ~ TRACTBNA * Callout
##   Resid. Df Resid. Dev Df Deviance
## 1        29     84.729
## 2         0      0.000 29   84.729
```

In this case, recalling the 5% threshold for a significant change in deviance is 42.6, and further noting that the 1% threshold is 49.6, there is strong evidence to reject a hypothesis of independence of rates between tracts, suggesting geographical variability in the two callout classifications. This is possibly because family disputes are likely to be located in residences, whereas breaches of the peace are more likely to occur over a wider range of places.

3.5.3 A further map-based example

An interesting issue for tidy data is what constitutes an 'observation'. Initially in the above example the callout was the *unit of observation* – details were stored about the kind of callout and its location. Thus each of these was a column in the sf table, and each row was a callout. However, in the Poisson regression, following the count transformation, a row was a count of a tract–callout-type combination – rather like a cell count in a cross-tabulation. The two rows were the tract and the callout type.

However, one thing suggested in the last example was that the geographical distribution of callouts for family disputes and breaches of the peace differed. It may be useful to map these two quantities – or the ratio between the two counts. One way of thinking about this is to refocus the unit of observation to be the tract. In this case, the table disturb_counts is essentially too 'gathered' – if the unit of analysis is the tract, then each tract should have a single row, with columns for the counts of each type of disturbance. Currently there is a single count column (n), a single Callout column stating which kind of callout is counted, and a tract column. Each tract ID (TRACTBNA) appears twice, once for each callout type. If the unit of observation is the tract, then tidy rules are violated, as a single observation is then occupying *two* rows. This can be transformed via the spread function. This is roughly the opposite of gather: it takes a column with stacked factors (or character strings) and a companion column with associated values, and spreads these across several columns, one for each of the stacked factor levels.

In the following code, spread is used to reshape the disturb_counts table in this way, printing out the result:

```
disturb_counts %>% spread(Callout,n)

## # A tibble: 31 x 3
##      TRACTBNA     `Breach of Peace`    `Family Dispute`
##      <fct>                  <int>               <int>
##   1 1401                      16                  10
##   2 1402                       5                  12
##   3 1403                      10                   9
##   4 1404                       1                   9
##   5 1405                       6                  13
##   6 1406                       9                  25
##   7 1407                      10                  17
##   8 1408                      11                  17
##   9 1409                       4                  15
## 10 1410                       0                  10
## # ... with 21 more rows
```

Next, the percentage of the sum of both attributable to family disputes is computed, and this is then left_joined to tracts:

```
disturb_counts %>%
    spread(Callout,n) %>%
    transmute(TRACTBNA,
              PC_DISPUTE=100*`Family Dispute`/
                  (`Family Dispute` + `Breach of Peace`)) %>%
    left_join(tracts,.) -> tracts2
```

The last line may look unfamiliar – the intention here is to pipeline the result of the transmute to a left join, but here we wish to make it the *second* argument (the first being the tracts sf table) because we want to join the outcome onto tracts to obtain an updated sf object – still a geographical table with polygons for the tracts and so on. The dot notation here causes that to happen – and generally it is a useful tool when non-standard pipelining is required.

As a result of this, tracts2 has a new column PC_DISPUTE giving the percentages of disturbance callouts that were family disputes (Figure 3.4):

```
tm_shape(tracts2) +
  tm_polygons(col='PC_DISPUTE',
              title='Family Disputes (%)') +
  tm_layout(legend.position = c('left','bottom'))
```

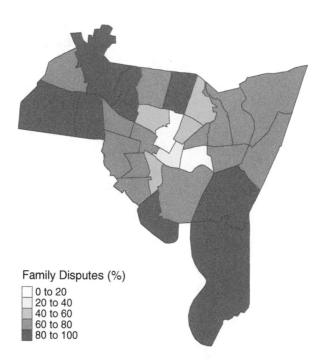

Figure 3.4 Family disputes as a percentage of disturbance callouts

Finally, an alternative view might be to consider each of the types of disturbance in separate maps showing the percentage of the total of each particular kind of callout associated with each tract. This is more in concordance with Tufte's notion of 'small multiples', as advocated in Tufte (1983, 1990).

This is relatively simple, requiring a slightly different calculation to compile the sf (here called tracts3):

```
disturb_counts %>%
  spread(Callout,n) %>%
  transmute(TRACTBNA,
            PC_FAM = 100*`Family Dispute`/sum(`Family Dispute`),
            PC_BOP = 100*`Breach of Peace`/sum(`Breach of Peace`)) %>%
  left_join(tracts,.) -> tracts3
```

This may then be shown as a pair of side-by-side maps (Figure 3.5):

```
tm_shape(tracts3) +
  tm_polygons(col=c("PC_FAM","PC_BOP"),
              title=c("Family Disputes (%)",
                      "Breach of Peace (%)")) +
  tm_layout(legend.position = c('left','bottom'))
```

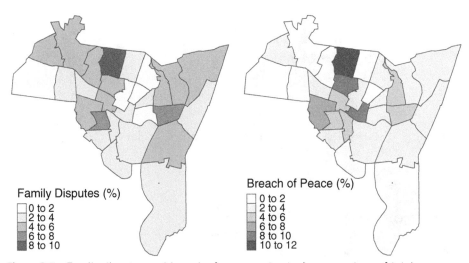

Figure 3.5 Family disputes and breach of peace as tract-wise percentage of total

Here it can be seen that although some tracts are foci for both kinds of callout, there are notable differences in spatial distribution. Exploring interactively (by running the above code chunk with `tmap_mode('view')` preceding) the central area is more of a focus for breaches of peace, while areas in the north-east and north-west see higher proportions of family disputes. One area (Newhallville) seems to be a focus for both kinds of callout.

3.5.4 Other spatial manipulations

As well as `st_join`, there are other kinds of manipulation of `sf` objects offered by the `sf` package. In this subsection, some of these will be demonstrated.

st_intersection for two sf objects provides a new row for every unique intersection of the first object with the second. Thus, if roads is intersected with tracts and a particular road intersects three tracts, there will be three rows created in the resulting sf object. The geometry column will contain the section of the road that intersects with the corresponding tract. All of the columns from both of the objects are in the resultant object – at the intersection of a row from roads and a row from tracts the values in the columns will be those from the particular road and tract that correspond to the associated intersection. Note, however, that this means that a repeated column name in both sf objects leads to an error. To avoid this, offending columns should either by dropped or renamed for one of the sf objects.

This carrying through of columns can be used, in conjunction with filter, to select out the part of the road network in a given tract. Here all of the roads in tract 1408 are selected and then mapped:

```
roads %>%
   st_intersection(tracts) %>%
   filter(TRACTBNA==1408) ->
   roads1408
tm_shape(tracts) + tm_borders() +
   tm_shape(roads1408) + tm_lines()
```

Buffering is also possible via the st_buffer function. Here the roads subset is buffered by 50 m:

```
roads1408 %>% st_buffer(50)
```

This can then be clipped (to find the parts of the buffer inside tract 1408). This is done by re-intersecting with tracts and selecting those whose TRACTBNA is still 1408. Noting that roads1408 also has a TRACTBNA column, this is dropped before carrying out the intersection (see above):

```
roads1408 %>%
   st_buffer(50) %>%
   select(-TRACTBNA) %>%
   st_intersection(tracts) %>%
   filter(TRACTBNA == 1408) ->
   bufroads1408
```

This may then be mapped as in Figure 3.6:

```
tm_shape(bufroads1408) + tm_fill() +
   tm_shape(roads1408) + tm_lines() +
   tm_shape(tracts) + tm_borders()
```

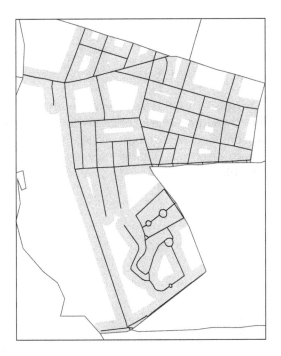

Figure 3.6   Road buffers clipped to a tract

The first tm_shape on any tmap construction determines the extent of the map – so that if subsequent shapes exceed these limits, only that part of them within the extent is drawn. This is useful here, allowing us to zoom in on the area where bufroads1408 is, thus highlighting detail.

Another useful spatial manipulation is achieved when applying summarise to sf objects. In this case, as well as applying a summary statistic to the grouping variable the shapes of observations in each group are 'unioned': that is, joined together with the removal of any internal boundaries. Below the wtd.var() function, weighted variance, from the Hmisc package (Harrell, 2018), is used in conjunction with sqrt to find the population weighted standard deviation of the variable P_OWNEROCC (percentage of owner-occupied residential properties) in each tract, based on block-level data. Note that the blocks sf object has these data, as well as a character column TRACTBNA stating which tract each block is in:

```
blocks %>% group_by(TRACTBNA) %>%
  summarise(OO_SD=sqrt(Hmisc::wtd.var(
    P_OWNEROCC, OCCUPIED, method="ML"))) ->
  tract_sd
```

The result may be mapped, giving an idea of which areas are more mixed in terms of owner occupation of residential properties:

```
tm_shape(tract_sd) +
  tm_polygons(col='OO_SD',
              title="Owner Occupation SD") +
  tm_layout(legend.position = c("left","bottom"))
```

A sometimes useful trick is to use this approach to unify all of the entities into a single sf. This is done by adding a new column with a constant value, grouping by this column, and summarising without any terms. Finally, if a boundary is required (i.e. a line-based object marking the edge of the study area, rather than a polygon), this can be achieved via st_boundary. A boundary object for blocks is created below:

```
tracts %>%
  mutate(gp=1) %>%
  group_by(gp) %>%
  summarise %>%
  st_boundary() -> bounds
```

A useful shorthand here is that it is possible to define grouping variables 'on the fly' in group_by – so the above can be replaced by:

```
tracts %>%
  group_by(gp=1) %>%
  summarise %>%
  st_boundary() -> bounds
```

This may now be used alongside the tracts and blocks simple features to show the different hierarchy is the spatial data in Figure 3.7:

```
tm_shape(bounds) + tm_lines(lwd=4,col='dodgerblue') +
tm_shape(tracts) + tm_borders(lwd=2,col='darkred') +
tm_shape(blocks) + tm_borders()
```

Key Points:

- Pipeline chaining and dplyr manipulation can be used with spatial data in sf format in the same way as a data.frame or tibble, for example to select variables (fields), filter rows (records) or mutate variables.

- The outputs of dplyr chaining can be piped directly to tmap mapping functions in the same way as for ggplot (see the end of Chapter 2 and Chapter 5).

- For statistical analysis of spatial data in `sf` format, the geometry may sometimes need to be removed using `select(-geometry)` or `st_drop_geometry()`.

- Functions in the `sf` package start with `st_`, and these can be used in `dplyr` piped operations, such as `st_join()`, `st_intersection()`, `st_boundary()` and `st_buffer()`.

Figure 3.7   Various levels of boundary for New Haven data

3.6 SUMMARY

In this chapter an integrated approach combining `dplyr` piping syntax and a tidy data approach to data manipulation and analysis has been described, later focusing on how these ideas can be applied to spatial information stored as `sf` objects.

There are a number of advantages and disadvantages to this approach – and the decision as to whether to adopt it may well depend on the particular problem in hand.

For a number of problems it provides a clear framework for the data structure used, and a number of packages (such as `tmap` and `ggplot`) are designed to work with tidy format data. Another advantage is that it is designed to work with data in a format that many databases can represent – and functions such as `left_join()`, `summarise` and `group_by` mirror very similar functions in SQL, used by database software (see, for example, Obe and Hsu, 2012), which will be discussed in detail in the following chapter. Indeed the `dbplyr` library (Wickham and Ruiz, 2018) allows `dplyr`-like functions to be applied to external database connections. Thus, most of the ideas here can be applied directly to large databases. Although the examples here were relatively small, very similar methods can be applied to very large datasets (big data) and spatial data.

One example of a situation where this format is perhaps less useful is in modelling spatial dependencies using techniques such as the spatial lag model or spatial error model (see, for example, Anselin, 2013) – here contiguity between geographical zones, such as tracts or blocks, is considered as part of the error model for a regression analysis and represented as a matrix. Although sparse contiguity information can be stored as tidy data, this may not be the most computationally effective way to proceed (given there could be a need to compute matrix inverses or eigenvectors). However, a well-designed package could perhaps allow the analysis to take place 'under the bonnet' so that the information output may appear to be in tidy form even if the actual calibration internally would not fit this model.

In conclusion, then, although there are some situations where the `tidyverse/dplyr/dbplyr` approach may not be the most suitable approach, there are many where it could well be, and therefore a good knowledge of the underpinning concepts is essential for spatial data analytics.

REFERENCES

Anselin, L. (2013) *Spatial Econometrics: Methods and Models* (4th edn). Dordrecht: Springer Science & Business Media.

Broman, K. W. and Woo, K. H. (2017) Data organization in spreadsheets. PeerJ Preprints 6:e3182. https://peerj.com/preprints/3183/.

Chatfield, C. (1995) *Problem Solving: A Statistician's Guide* (2nd edn). London: Chapman & Hall.

Harrell Jr, F. E. (2018) *Hmisc: Harrell Miscellaneous*. https://CRAN.R-project.org/package=Hmisc.

Obe, R. and Hsu, L. (2012) *PostgreSQL: Up and Running*. Sebastopol, CA: O'Reilly Media.

Tufte, E. R. (1983) *The Visual Display of Quantitative Information*. Cheshire, CT: Graphics Press.

Tufte, E. R. (1990) *Envisioning Information*. Cheshire, CT: Graphics Press.

Wickham, H. and Ruiz, E. (2018) *Dbplyr: A 'Dplyr' Back End for Databases*. https://CRAN.R-project.org/package=dbplyr.

4

CREATING DATABASES AND QUERIES IN R

4.1 OVERVIEW

Data can be so voluminous (big) that they are difficult to work with on a standard computer: they may not fit on hard drive or disk if they are terabytes in size, and even if they are smaller, the computer may struggle to hold them in working memory and in the memory allocated to the program you are using (including R).

For these reasons, we increasingly work with data that are held *elsewhere* in some kind of database system that we access remotely. Queries are sent to the database and the results are returned. Such queries can be formulated to extract and return *only* the data we are interested in, usually a (much) smaller subset of the data, which has been manipulated and summarised in some way. Queries can also be used to link data in different database tables before extracting, grouping or summarising. The key thing is that the data in their raw unqueried form are not in the working memory of our computers.

This chapter provides an introduction to databases and why we need them. It covers the main ideas behind database management systems, relational databases and database functionality in R and describes how to set up in-memory databases and local ones. The use of `dplyr` functions (verbs and joins) to interrogate or *query* databases is illustrated, emphasising the inherently *lazy* nature of `dplyr` queries (i.e. such that the data remain on the database for as long as possible rather than being pulled into working memory). A query to a large database of medical prescriptions, combining and manipulating data from multiple tables, is developed and the results will be used in later chapters (Chapters 5 and 6) to illustrate what we believe to be core concepts in data analytics centred around the need to understand data structure prior to analysis.

For this chapter you will need the following packages installed and loaded:

```
library(stringr)
library(tidyverse)
library(DBI)
library(tmap)
```

You will also need to undertake some data downloads, one of which is 11 GB and will take some time.

4.2 INTRODUCTION TO DATABASES

4.2.1 Why use a database?

Up until now in this book all of the code snippets have been applied to *in-memory* data: the data were loaded into R in some way, read from either an external file or one of R's internal datasets. These data have been of a small enough size for R to manage them in its internal memory. For example, the lsoa_data in Chapter 2 was 3.2 MB and the newhaven dataset in Chapter was 3.0 MB. The main reason for using a database is that that data are too big and cumbersome for easy manipulation in the working memory of whatever analysis environment you are working in (R, Python, QGIS, etc.). Essentially it is for data storage and access considerations.

Consider a flat 10 GB data table. Any analysis will typically want to select certain rows or columns based on specific criteria. Such operations require a sequential read across the entire dataset. Reading (and writing) operations take time. For example, if reading 1 MB sequentially from memory takes 0.2 seconds (0.005 seconds from a solid state disk) then to access an in-memory 10 GB file would take 50 seconds, and accessing a file from disk would be about 10 times slower. If the file is modified and has to be written, then this takes longer. If the file is very large (e.g. 100 GB) then such rates would be optimistic as they assume that all the data can be read in a single continuous read operation. Typically, disk storage is increased and memory remains a limiting factor as not everything can be fit into memory. Additionally, any kind of selection or filtering of data rows or columns requires linear searches of the data, and although these can be quicker if the data are structured in some way (e.g. by sorting some or all of the columns), there is a trade-off between storage and efficiency. The problem is that sorting columns is expensive as sorted columns (or *indexes*) require additional storage. Thus because of the slowness of reading and writing, data are often held in indexed databases that are remotely stored (i.e. they are neither on your computer nor in your computer's memory) and accessed.

The basic idea behind databases (as opposed to data frames, tibbles and other *in-session* data table formats) is that you connect to a database (either a local one, one in working memory or a remote one), compile queries that are passed to the database, and only the query result is returned.

Databases are a collection of data tables that are accessed through some kind of database management system (DBMS) providing a structure that supports database queries. Databases frequently hold multiple data tables which have some field (attribute) in common supporting *relational* queries. Records (observations) in different data tables are *related* to each other using the field they have in common (e.g. postcode, national insurance number). This allows data in different tables to be combined, extracted and/or summarised in some way through *queries*. Queries in this context are specific combinations of instructions to the DBMS in order to retrieve data from the server.

In contrast to *spatial* queries (see Chapter 7 and other examples throughout this book) which use some kind of spatial test based on a topological query (within a certain distance, within a census area, etc.) to link data, relational queries link data in different tables based on common attributes or fields.

4.2.2 Databases in R

Relational databases and DBMSs classically use Structured Query Language (SQL) to retrieve data through queries. SQL has a relatively simple syntax (supported by ISO standards) but complex queries can be difficult to code correctly. For these reasons the team behind `dplyr` constructed tools and functions that translate into SQL when applied to databases. Workflows combine the various `dplyr` verbs used for single-table and two-table manipulations (Chapters 2 and 3), translate them into SQL and pass them on to the DBMS.

To query databases in this way requires a database interface (DBI). Fortunately, a DBI in the form of the `DBI` package is installed with `dplyr`. The DBI separates the connectivity to the DBMS into a *front-end* and a *back-end*. The front-end is you or your R session and the analyses and manipulations you wish to undertake, and the back-end is the database and the queries that are passed to it. Applications such as `dplyr` use only the exposed *front-end* of the DBMS and allow you to connect to and disconnect from the DBMS, create and execute queries, extract the query results, and so on. The `dbplyr` package contains back-end tools that translate the `dplyr` functions, calls and queries into SQL in order for them to work with remote database tables, with the DBI defining an interface for communication between R and relational DBMSs. So you have used `dbplyr` before without knowing it in Chapter 3. If you look at the help you will see that the `dbplyr` package contains only a small number of wrapper functions – but there are many `dplyr` functions that use `dbplyr` functionality in the background. Thus `dbplyr` is used when, for example, you use the `select` function in `dplyr`.

This book uses SQLite because it is embedded inside an R package (`RSQLite`) which is automatically loaded with `dplyr`. This provides a convenient tool for understanding how to manage large datasets because it is completely embedded inside an R package and you do not need to connect to a separate database server. A good introduction to this topic can be found in Horton et al. (2015). The `DBI` package provides an interface to many different database packages. This allows you to use the same R code, for example with `dplyr` verbs, to access and connect to a number of back-end database formats including MySQL with the `RMySQL` package and Postgres with `RPostgreSQL`.

4.2.3 Prescribing data

A remote database of around 121 million medical prescriptions issued in England in 2018 by some 10,600 GP practices called `prescribing.sqlite` will be used to undertake some spatial analyses in Section 4.5. However, as this is 11.2 GB in size, a smaller version with a sample of 100,000 prescriptions has been created to illustrate how to set up databases and to query them. This will be used in the code below to create `prescribing_lite.sqlite`.

The sampled data tables are contained in an .RData file called ch4_db.RData which can be downloaded and saved to a local folder. The code below loads the file from the internet to your current working directory:

```
# check your current working directory
getwd()
download.file("http://archive.researchdata.leeds.ac.uk/732/1/
          ch4_db.Rdata",
          "./ch4_db.RData", mode = "wb")
```

Now load the ch4_db.RData file to your R/RStudio session and examine the result:

```
load("ch4_db.RData")
ls()
## [1] "lsoa_sf"        "patients"    "postcodes"    "practices"
## [5] "prescriptions"  "social"
```

You will notice that five data tables and one sf spatial data object are loaded to your R session. The data tables are prescriptions, practices, postcodes, patients and social. You should examine these – as a reminder the functions str, summary and head are useful for this, as is the casting of the data.frame to a tibble format using as_tibble:

```
str(prescriptions)
head(practices)
as_tibble(postcodes)
summary(social)
head(patients)
lsoa_sf
```

The prescriptions data table records the prescriptions issued by each general practice (doctor's surgery) and includes the British National Formulary (BNF) code of each prescription, its name, the number of items prescribed, their cost and quantity. It indicates the month they were prescribed and also includes an individual ID for the GP or doctor's practice and the health authority or primary care trust in which the practice sits. It contains no information about individual patients or doctors.

The `practices` data table contains the details for each GP practice, including its ID and address. The `prescriptions` and `practices` data tables were downloaded from the UK government's Open Data portal (https://data.gov.uk/dataset/176ae264-2484-4afe-a297-d51798eb8228/gp-practice-prescribing-data-presentation-level).

Along with this, the numbers of patients from each census area (Lower Super Output Areas, LSOAs) for each GP practice are contained in the `patients` data table, in long format. This lists the GP practice codes (`practice_id`), an LSOA identifier (`lsoa_id`) and the count of patients from each GP practice in each LSOA area (from http://digital.nhs.uk/catalogue/PUB22008).

The `postcode` data table was downloaded from the Office for National Statistics (http://geoportal.statistics.gov.uk/datasets/dfa0ff74981b4b228d2030d852f0b14a) and for each postcode it includes the easting and northing (OSGB projected coordinates) and the LSOA code in which the postcode centre sits.

The `social` data table describes the following UK 2011 population census attributes for each LSOA: the resident population (`pop`), the proportions of the population that are employed, unemployed, have a limiting long term illness (`llti`), lack qualifications (`noqual`), have post-16 qualifications (`14qual`), work fewer than 15 hours per week (`ptlt15`), work 16–30 hours (`pt1630`), work 31–48 hours (`ft3148`) and work 49 or more hours (`ft49`). It also includes the Rural Urban Classification (see https://www.gov.uk/government/statistics/2011-rural-urban-classification) and a geo-demographic classification code for the LSOA from the Output Area Classification (http://www.opengeodemographics.com). UK Census data can be downloaded from the Office for National Statistics (https://www.nomisweb.co.uk).

The `lsoa_sf` spatial dataset contains the outlines of the LSOAs in England and Wales. Census areas for the UK can be downloaded directly from the UK Data Service (https://borders.ukdataservice.ac.uk).

The attributes in each data table and the links (relations) between the data tables are shown in Figure 4.1 (note that the `lsoa_id` can also be used to link to the `lsoa_sf` spatial data). The figure shows the pairs of relationships between data tables, and critically their cardinality. For the `prescribing.sqlite` database:

- `prescriptions` connects to `practices` via a single variable, `practice_id`;

- `practices` connects to `postcodes` through the `postcode` variable;

- `postcodes` connects to `social` via the `lsoa_id` variable.

These relationships suggest how a chain of relational queries could be specified – for example, to link the volume of prescriptions of a particular type to the population in LSOA census areas, in order to determine a rate or cost of prescribing per head of population, and to examine how this relates to socio-economic variables in the `social` data table. Rowlingson et al. (2013) provide an overview of the data and how they can be used to develop a spatial analysis of prescribing.

Key Points

- Very large data files can be problematic to work with on standard computers.

- For these reasons we often use data in data tables stored elsewhere in databases that we access remotely.

- The data tables held in a *relational* database are linked through common fields (attributes) and are accessed through a database management system that provides a structure for supporting database queries.

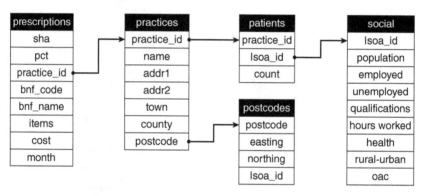

Figure 4.1   The relations between the prescribing data tables

- Many of the functions in the dplyr package can be used to query databases. An R version of a database querying engine, SQLite (RSQLite), is automatically loaded with dplyr.

- Two prescribing databases were introduced (one a sample of 100,000 prescriptions, the other with 121 million records).

4.3 CREATING RELATIONAL DATABASES IN R

The basic idea for creating databases is that you define or open a connection from your R/RStudio session to a named database. The database is populated with data and then you can work with it. The connection should be closed when you finish working with the database.

There a number of different types of databases that can be created:

- Local *in-memory* databases are useful for prototyping and testing. They are temporary and are automatically deleted when you disconnect from them.

- Local *on-file* databases are permanent databases held locally (i.e. on your computer).

- Remote *on-file* databases are permanent and held elsewhere (i.e. on another computer!), and require an internet connection to access them. The creation of these is not covered here, but once you are connected to the remote server, the same operations can be undertaken.

4.3.1 Creating a local in-memory database

Local *in-memory* databases are useful for illustrating the principles of database creation and of constructing queries. In reality, if the data fit into working memory and can be manipulated and accessed ('wrangled') with ease, then it is likely that you do not need to use databases and queries: standard R data formats such the data. frame and tibble and related operations with dplyr will probably suffice.

To work with a database in dplyr, the connection to it needs to be specified. This is done with the DBI::dbConnect() function. This can be interpreted as saying 'use the Connect() function from the DBI package' – the :: indicates the R package to which the function belongs (recall that the DBI package is loaded with dbdplyr, which in turn is loaded by dplyr).

The code snippet below defines an in-memory database:

```
db = DBI::dbConnect(RSQLite::SQLite(), path = ":memory:")
```

The arguments to DBI::dbConnect() vary from database to database, but the first argument is always the database back-end. It is RSQLite::SQLite() for RSQLite, RPostgreSQL::PostgreSQL() for RPostgreSQL, etc. The SQLite() implementation in R (RSQLite) only needs one other argument: the path to the database. Here we use the special string "memory" which causes SQLite to make a temporary in-memory database (i.e. in the working memory of R/RStudio).

The database, db, has no data in it. To populate the database, the dplyr function copy_to can be used. This uploads data to the database. The code below does this for the prescriptions data table as an example:

```
copy_to(db, prescriptions, name = "prescripts_db")
```

The tbl function makes the connection to the data table:

```
tbl(db, "prescripts_db")
```

The above code snippet copied *all* of the data to db, but it is also possible to specify particular fields. The code below overwrites prescriptions in db using the dbWriteTable() function in the DBI package:

```
dbWriteTable(conn = db, name = "prescripts_db",
    value = prescriptions[, c("sha","bnf_code","act_cost","month")],
    row.names = FALSE, header = TRUE, overwrite = T)
```

```
tbl(db, "prescripts_db")

## # Source:    table<prescripts_db> [?? x 4]
## # Database: sqlite 3.30.1 []
##     sha    bnf_code          act_cost month
##     <chr>  <chr>                <dbl>  <chr>
##  1  Q48    131002030AAADAD        5.3  04
##  2  Q56    0204000HOBFADAA       16.4  05
##  3  Q66    1001010AJAAADAD      104.   07
##  4  Q67    0407020B0BHACAJ      240.   11
##  5  Q55    1001010AHBBAAAA       20.0  03
##  6  Q63    1001010J0AAAEAE       44.4  11
##  7  Q53    0202030X0AAAAAA       18.7  12
##  8  Q68    0906040G0AAEKEK        2.56 08
##  9  Q57    130201100BBBAAN        6.61 08
## 10  Q63    0205051R0AAALAL       58.6  01
## # ... with more rows
```

In fact copy_to is a wrapper for dbWriteTable and the same operation can be undertaken with:

```
copy_to(db,
        prescriptions[, c("sha", "bnf_code", "act_cost","month")],
        name = "prescripts_db", overwrite = T)
```

The tbl() function is very useful here as it allows us to extract data from the database and to pipe it around (i.e. do things with it). The way that tbl works is to create a reference to the data table in the database (e.g. prescripts_db). An alternative is to use dbReadTable to do this, but it pulls the data from the database to working memory. This is illustrated using the dim() (dimensions) function in the code below:

```
dim(as_tibble(dbReadTable(db, "prescripts_db")))
## [1] 100000      4
dim(tbl(db, "prescripts_db"))
## [1] NA 4
```

The dplyr piping syntax can be applied in conjunction with tbl to select, filter manipulate, summarise and extract data from the database. The example below selects records from the prescripts_db data table in db that are in Strategic Health Authority (SHA) Q49 (the sha field in the data) and cost more than £100:

```
tbl(db, "prescripts_db") %>%
  filter(sha == "Q49" & act_cost > 100)
```

A similar syntax can be used to work out average prescription costs for different SHAs:

```
tbl(db, "prescripts_db") %>%
    group_by(sha) %>%
    summarise(mean_cost = mean(act_cost, na.rm = T))
```

This starts to suggest how queries can be constructed.

When we have finished with any database connection we have to close it using the dbDisconnect function:

```
dbDisconnect(db)
```

4.3.2 Creating a local on-file database

The in-memory database ceases to exist when the connection is closed. An *on-file* database is permanent and, after being populated with data, can be connected to in later R sessions. Here the data tables contained in the ch4_db.RData file are used to create an RSQLite database called prescribing_lite.sqlite. This will contain the five data tables and the lsoa_sf spatial data. The process is similar to creating an in-memory database, but this time a database file is created.

Note that you may wish to create a working directory for data you create that is distinct from the one you use to download data. The code below creates a data folder called data.db in your current working directory that could be used to hold the database we are about to create and sets the working directory to this folder:

```
dir.create("./data.db")
setwd("./data.db")
```

In the code below, the first line checks for the existence of prescribing_lite. sqlite and removes it if it exists. The second then creates a connection to the database:

```
if (file.exists("prescribing_lite.sqlite") == TRUE)
    file.remove("prescribing_lite.sqlite")
db = DBI::dbConnect(RSQLite::SQLite(),
                    dbname="prescribing_lite.sqlite")
```

If you look in your working folder using Windows Explorer (PC), your file manager in Linux or Finder (Mac), you will see that an empty object called prescribing_lite.sqlite has been created. It is empty and has no size because it has not yet been populated with data.

The following code populates the db object with the tables in Figure 4.1 using the dbWriteTable function, and then closes the connection to the database:

```
dbWriteTable(conn = db, name = "prescriptions",
             value = prescriptions,
             row.names = FALSE, header = TRUE)
dbWriteTable(conn = db, name = "practices", value = practices,
             row.names = FALSE, header = TRUE)
dbWriteTable(conn = db, name = "patients", value = patients,
             row.names = FALSE, header = TRUE)
dbWriteTable(conn = db, name = "postcodes", value = postcodes,
             row.names = FALSE, header = TRUE)
dbWriteTable(conn = db, name = "social", value = social,
             row.names = FALSE, header = TRUE)
dbDisconnect(db)
```

Again, note the use of the dbDisconnect function in the last line. If you enter db at the console you will see that it is disconnected.

If you check again in your working folder you will see that prescribing_lite. sqlite now has been populated and has a size of around 35 MB.

To access the data, we need simply to connect to the database, having created it:

```
db <- DBI::dbConnect(RSQLite::SQLite(),
                     dbname="prescribing_lite.sqlite")
```

You can check what the database contains using different commands from the DBI package:

```
# tables in the database
dbListTables(db)

## [1] "patients"      "postcodes"      "practices"
## [4] "prescriptions"  "social"

# fields in the table
dbListFields(db, "postcodes")
## [1] "postcode" "easting" "northing" "lsoa_id"
```

It is possible to query the data tables in the database. For example, the code below summarises the mean cost of prescriptions in each of the SHAs, arranges them in descending mean cost order, filters them for at least 100 items and prints the top 10 to the console:

```
tbl(db, "prescriptions") %>%
  group_by(sha) %>%
  summarise(
    mean_cost = mean(act_cost, na.rm = T),
    n = n()
  ) %>%
```

```
ungroup() %>%
arrange(desc(mean_cost)) %>%
filter(n > 100) %>% print(n=10)
```

However, this is working with only one of the data tables in the db database. It is not taking advantage of the layered analysis that is possible by linking data tables, as shown in Figure 4.1.

To link the data tables in the db database we need to specify the links between them, just as we did using the join functions in Chapter 2. However, to join in the right way, we need to think about how we wish to construct our queries and what any queries to database tables will return (and therefore the analysis and inference they will support). The DBI package has a number of functions that list data tables and their fields as above. Similarly, tbl in the dplyr package can be used to do these summaries:

```
tbl(db, "social")
tbl(db, "practices")
tbl(db, "patients")
colnames(tbl(db, "prescriptions"))
```

The results are similar to the str, head and summary functions used earlier. However, it is important to restate what the tbl function is doing: it is sending a tbl query to db about the table (e.g. the prescriptions table) in db. As tbl returns the first 10 records this is what is returned from db – the data returned to the R session are only the result of this query and not the whole of the data table. To emphasise this, the code below compares the size of the query and the size of the prescriptions data table that is loaded into the R session (note that there are small differences in the way that file sizes are calculated between Windows and Mac operating systems):

```
format(object.size(tbl(db, "prescriptions")), unit="auto")
## [1] "5.3 Kb"

format(object.size(prescriptions), unit="auto")
## [1] "11.2 Mb"
```

Again the database can be closed as follows:

```
dbDisconnect(db)
```

4.3.3 Summary

The basic process of creating a database is to first define a connection to a database, then populate it with data, and, when it is populated, to close the connection.

Two kinds of database were illustrated: an in-memory one, good for development; and a local on-file database. The procedures for constructing these were essentially the same:

1. Define a connection to a database.

2. Populate the database with data.

3. Close the database.

This sequence can be used to construct much larger databases.

Key Points

- Different types of databases can be created including local in-memory databases for prototyping and testing, local permanent on-file ones, and remote databases held elsewhere on another computer.

- The basic process of creating a database is first to define a connection to a database, then to populate it with data, and, when it is populated, to close the connection.

- Once instantiated or connected to, the `copy_to` function in `dplyr` or the `dbWriteTable` function in the `DBI` package can be used to populate the database tables with data.

- The main principle in working with databases is that the data reside elsewhere (i.e. not in working memory) and only the results of queries are pulled from the database, not the data themselves.

- The `tbl` function in `dplyr` creates a reference to the data table in the database without pulling any data from the database, allowing subsequent queries to be constructed.

4.4 DATABASE QUERIES

In this section we will explore a number of core database operations using `dplyr` to construct queries and apply them to the `prescribing_lite.sqlite` database. There are three main groups of operations that are commonly used in database queries, either singly or in combination. Queries specify operations that:

1. Extract (specifying criteria, logical pattern matching, etc.)

2. Join (linking different tables)

3. Summarise (grouping, using summary functions, maintaining fields, creating new fields).

There are some overlaps with Sections 2.4 and 3.3 in terms of the operations using dplyr tools, but here these are applied in a database context. In many cases the syntax is exactly the same as those applied to data.frame and tibble formats in the previous chapters. On occasion, however, they need to be adapted for working with databases.

4.4.1 Extracting from a database

It is possible to extract whole records (rows) and fields (columns), individual rows and columns that match some criteria, and individual elements (cells in a data table). The two most commonly used approaches for extracting data are:

1. By specifying some kind of logical test or conditions that have to be satisfied for the data to be extracted

2. By specifying the location of the data you wish to extract, for example by using the *i*th row or *j*th column, or variable names.

Logical queries have TRUE or FALSE answers and use logical operators (e.g. greater than, less than, equal to, not equal to). These have been covered in previous chapters, but the full set of logical operators can be found in the R help (enter ?base::Logic). The other main way of selecting is through some kind of text pattern matching. Both may be used to subset database fields (columns) by filtering and/or database records (rows) by selecting.

You should be familiar with the dplr *verbs* introduced in Chapter 2 and illustrated in Chapter 3. These can be applied to databases in the same way.

A connection has to be made for a database to be queried. Connect to the prescribing_lite.sqlite database you created earlier as before:

```
library(RSQLite)
db <- dbConnect(SQLite(), dbname="prescribing_lite.sqlite")
```

The code below uses filter() to extract the prescriptions for a specific flu vaccine via the BNF code (see, for example, https://openprescribing.net/bnf/) and orders the extracted records by volume (items):

```
tbl(db, "prescriptions") %>%
  filter(bnf_code == "1404000H0AAAFAF") %>%
  arrange(desc(items))
```

For multiple matches, the %in% function can be used:

```
tbl(db, "prescriptions") %>%
  filter(bnf_code %in% c("1404000H0AAAFAF", "1305020D0BEAAAI")) %>%
  arrange(desc(items))
```

The logical statement above using == returns *exact* matches. Now we might want to filter for all the flu vaccine prescriptions, which start with the BNF code 1404000H0. In Chapter 3 the str_detect function was used to filter for callouts that involved the term 'burglary'. However, the stringr pattern matching functions within dplyr, all in the form str_<action>, do not work with databases.

For filtering operations, %like% can be applied:

```
# one pattern
tbl(db, "prescriptions") %>%
   filter(bnf_code %like% '%1404000H0%')
# multiple patterns
tbl(db, "prescriptions") %>%
   filter(bnf_name %like% 'Dermol%' |
             bnf_name %like% 'Fluad%')
```

However, pattern matching in this way needs to be done with *extreme care* because it undertakes *partial matching*, whereas filtering using logical approaches with == or %in% will only return exact matches. Additionally, the code above includes some wildcards \% at the start and end of the pattern passed to %like%. This means that the filter operation will return any records that have bnf_code values containing 1404000H0, not just those that start with 1404000H0.

A final filtering consideration is selecting by row number, often after some kind of ordering. The slice function works with data tables loaded into the R session but not with database connections, so row_number can be used as below:

```
tb.l(db, "prescriptions") %>%
   filter(bnf_code == "1404000H0AAAFAF") %>%
   arrange(desc(items)) %>%
   filter(between(row_number(), 1, 5))
```

The select function can be used to return specific fields from a database in a similar way to working data tables loaded into the R session:

```
tbl(db, "social") %>%
   select(unemployed, llti)
```

Fields can also be selected using pattern matching, but here a set of tidy matching functions *can* be applied:

```
tbl(db, "prescriptions") %>%
   select(starts_with("bnf_"))
tbl(db, "prescriptions") %>%
   select(contains("bnf_"))
```

You should examine the help for these:

```
?tidyselect::select_helpers
```

Additional selection functions include row_num which extracts the records numerically:

```
tbl(db, "prescriptions") %>%
    select(num_range = c(1,3))
```

Thus far, all of the results of the code snippets applied to the database have been printed out to the console. No data have been returned to the R session, meaning that all of the analysis has taken place away from the working memory of your computer. The use of collect() returns the results. It creates an object (even if it is just printed out). To show this, run the code below which uses the object.size function to evaluate the memory cost to R of running the code snippets. This is a key advantage of working with databases.

```
# size of the call
tbl(db, "prescriptions") %>% object.size()
# size of a longer call
tbl(db, "prescriptions") %>%
    filter(bnf_code %like% '%1404%') %>%
    arrange(desc(act_cost)) %>% object.size()
# size of what is returned with collect
tbl(db, "prescriptions") %>%
    filter(bnf_code %like% '%1404%') %>%
    arrange(desc(act_cost)) %>% collect() %>% object.size()
```

4.4.2 Joining (linking) database tables

Chapter 2 introduced the idea of joins between different tables. Some (but not all) of the two-table dplyr joins can also be applied to tables held in databases. Figure 4.1 shows the links between the five data tables in the prescribing_lite.sqlite database, with each link provided through a common field or attribute, and with the different dplyr joins enforcing different types of cardinality.

These can be applied to the database in a similar way to working with data tables loaded into the R session. The code below filters for antidepressants, joins the results to data on the practices and then the postcodes database tables, before selecting the census area, cost and items fields:

```
tbl(db, "prescriptions") %>%
  filter(bnf_code %like% '%1404000H0%') %>%
  inner_join(tbl(db, "practices")) %>%
  inner_join(tbl(db, "postcodes")) %>%
  select(lsoa_id, act_cost, items)

## # Source:     lazy query [?? x 3]
## # Database: sqlite 3.30.1
## #    [/Users/geoaco/Dropbox/RDataSci/prescribing_lite.sqlite]
##      lsoa_id      act_cost  items
##      <chr>            <dbl>  <int>
##   1 E01016066      754.         83
##   2 E01014325      244.         40
##   3 E01008154        6.12        1
##   4 E01008636      757.        102
##   5 E01024689     3434.        378
##   6 E01028294     2227.        300
##   7 E01003450      831.        112
##   8 E01010372      269.         44
##   9 E01029034    10763.       1186
## 10 E01002561      507.         83
## # … with more rows
```

Note again that without the collect() function in the code above, the joined data remain in the database (i.e. are not returned to the console).

Also note that not all dplyr joins currently support database queries, although they may in the future:

```
tbl(db, "prescriptions") %>%
  full_join(tbl(db, "practices")) %>% collect() %>% dim()
tbl(db, "prescriptions") %>%
  right_join(tbl(db, "practices")) %>% collect() %>% dim()
```

This means that queries with nested calls to different database tables have to be carefully designed.

Of course, the dplyr verbs can also be applied, for example to filter the data as was done in the first join example here. The code below selects for antidepressants, undertakes two inner joins and specifies a set of attributes to be returned, in this case the SHA, month, the geographic coordinates of the practice, the drug code and the item cost. You should note the way that the code *forces* R to take the dplyr version of the select function, using the format package_name::fucntion_name. Sometimes you may have similarly named functions from different packages loaded in your session – the raster package, for example, also has a function called select.

```
tbl(db, "prescriptions") %>%
  filter(bnf_code %like% '%1404000H0%') %>%
  inner_join(tbl(db, "practices")) %>%
  inner_join(tbl(db, "postcodes")) %>%
  dplyr::select(sha, month, easting, northing, bnf_code, act_cost)

## # Source:    lazy query [?? x 6]
## # Database: sqlite 3.30.1
## #   [/Users/geoaco/Dropbox/RDataSci/prescribing_lite.sqlite]
##     sha   month  easting  northing bnf_code          act_cost
##     <chr> <chr>   <int>     <int>  <chr>                <dbl>
## 1   Q67   09     576493    165385  1404000H0BXAAAF       754.
## 2   Q60   01     389764    348721  1404000H0AAAFAF       244.
## 3   Q51   10     433202    388493  1404000H0AAAKAK        6.12
## 4   Q49   12     438490    565260  1404000H0BWAAAF       757.
## 5   Q67   09     638101    168201  1404000H0BXAAAF      3434.
## 6   Q55   10     480206    353539  1404000H0BWAAAF      2227.
## 7   Q63   11     526079    167182  1404000H0BWAAAK       831.
## 8   Q54   02     401501    297463  1404000H0AAAFAF       269.
## 9   Q65   11     378345    148728  1404000H0BXAAAF     10763.
## 10  Q62   01     509130    173409  1404000H0AAAFAF       507.
## # ... with more rows
```

4.4.3 Mutating, grouping and summarising

The final set of considerations in this section concerns methods for mutating, grouping and summarising data. The two most commonly used verbs to do this are mutate and summarise; the latter is implicitly linked to grouping functions, the most common of which is group_by.

The summarise() function summarises existing variables by the function that is passed to it. If the data are grouped it will return a summary for each group, otherwise it will summarise over the whole dataset. To illustrate these the code below summarises costs in the prescription table over the whole dataset and then groups these by sha, and by sha and month:

```
# entire dataset
tbl(db, "prescriptions") %>%
  summarise(total = sum(act_cost, na.rm = T))
# grouped by sha
tbl(db, "prescriptions") %>%
  group_by(sha) %>%
  summarise(total = sum(act_cost, na.rm = T)) %>%
  arrange(desc(total))
```

```
# grouped by sha and month
tbl(db, "prescriptions") %>%
  group_by(sha, month) %>%
    summarise(total = sum(act_cost, na.rm = T)) %>%
    arrange(desc(total))
```

Recall that summarise only returns variables created by its summary operations along with any grouping.

The mutate() function adds new variables for all the rows in the input data table and returns all variables present when it is applied. The code below summarises the costs of all prescriptions per practice and uses mutate to calculates the cost per item:

```
tbl(db, "prescriptions") %>%
  group_by(practice_id) %>%
    summarise(
      cost = sum(act_cost, na.rm = T),
      n = n()) %>%
    mutate(mean_cost = cost/n) %>%
    arrange(desc(mean_cost))
## # Source:      lazy query [?? x 4]
## # Database:    sqlite 3.30.1
## #    [/Users/geoaco/Dropbox/RDataSci/prescribing_lite.sqlite]
## # Ordered by: desc(mean_cost)
##      practice_id   cost      n mean_cost
##      <chr>        <dbl> <int>     <dbl>
##   1 Y04558       12030.     2      6015.
##   2 Y03488        4805.     1      4805.
##   3 Y05297        4289.     1      4289.
##   4 Y05351        1492.     1      1492.
##   5 Y03739        1430.     1      1430.
##   6 N82667        1295.     1      1295.
##   7 L82009       18707.    16      1169.
##   8 Y04223        1974.     2       987.
##   9 G82095       15491.    18       861.
## 10 Y05017        4902.     6       817.
## # ... with more rows
```

This operation can be unpicked by running the query in discrete steps and examining the intermediate outputs. First the grouping and the summary:

```
tbl(db, "prescriptions") %>%
  group_by(practice_id) %>%
    summarise(
      cost = sum(act_cost, na.rm = T),
      n = n())
```

Then the creation of a new variable with mutate:

```
tbl(db, "prescriptions") %>%
  group_by(practice_id) %>%
  summarise(
    cost = sum(act_cost, na.rm = T),
    n = n()) %>%
  mutate(mean_cost = cost/n)
```

Then the final ordering by adding the pipe and arrange(desc(mean_cost)) as above. And the observant among you will notice that this is the mean!

```
tbl(db, "prescriptions") %>%
  group_by(practice_id) %>%
  summarise(mean_cost = mean(act_cost, na.rm = T))%>%
  arrange(desc(mean_cost))
```

4.4.4 Final observations

This section has described the different kinds of operations for constructing database queries using the dplyr syntax. These include:

- functions for extracting data (rows/columns) from a database based on pattern matching, logical tests and specific positional references;

- functions for joining or linking database tables together based on some common property or attribute, with consideration of the cardinality of the link, applying the joins first introduced in Chapter 2;

- functions for summarising data, with or without grouping, applying different functions applying the dplyr verbs, also introduced in Chapter 2.

These can be used to construct complex queries (a worked example is given in the next section) of filtered records, selected fields and joined database tables. Using dplyr to do this has a number of advantages, the main one being that dplyr tries to be *lazy* by never pulling data into R's memory unless explicitly requested (e.g. by using the collect function). This is done because dplyr queries create *references* to the data in the database, and the results that are returned (printed) to the console are just summaries of the query. When collect() is used the dplyr query results are retrieved to a local tibble.

However, in some cases we may want to pull the data down from the database and perform our database operations. The dbReadTable() function does this and returns similar results, but operates in a very different way, as indicated by the size of the R objects they create and different times they take to run:

```
object.size(tbl(db, "prescriptions"))
object.size(dbReadTable(db, "prescriptions"))
```

One other aspect of dplyr is that it compiles all of the database commands and translates them into SQL before passing them to the database in one step. This can be illustrated by adding show_query() to the code snippet above:

```
tbl(db, "prescriptions") %>%
  group_by(practice_id) %>%
    summarise(mean_cost = mean(act_cost, na.rm = T))%>%
    arrange(desc(mean_cost)) %>%
    show_query()

## <SQL>
## SELECT `practice_id`, AVG(`act_cost`) AS `mean_cost`
## FROM `prescriptions`
## GROUP BY `practice_id`
## ORDER BY `mean_cost` DESC
```

It is also possible to pass SQL code directly to the database using the dbGetQuery function in the DBI package:

```
dbGetQuery(db,
  "SELECT `practice_id`, AVG(`act_cost`) AS `mean_cost`
  FROM `prescriptions`
  GROUP BY `practice_id`
  ORDER BY `mean_cost` DESC") %>% as_tibble()
```

SQL is a very powerful language with a standard syntax. Further information on constructing SQL queries can be found at https://www.sqlite.org/queryplanner.html. The database can be closed as follows:

```
dbDisconnect(db)
```

Key Points

- Database queries are used to select, filter, join and summarise data from databases.

- Selection and filtering are undertaken using logical statements or by specifying the named or numbered location (records, fields) of the data to be extracted.

- Queries can be constructed using dplyr syntax for data manipulations, selection, filtering, mutations, joins, grouping and so on, which generally use the same syntax as `dplyr` operations on in-memory data tables described in Chapter 3.

- There are some differences, especially with pattern matching (e.g. for multiple pattern matches, the `%in%` function can be used).

- `dplyr` translates queries into SQL which is passed to the database and the SQL code generated by the `dplyr` query can be printed out with the `show_query()` function.

- SQL code can be passed directly to the database using the `dbGetQuery` function in the `DBI` package.

- The `collect()` function at the end of a `dplyr` query returns the query result to the current R session.

4.5 WORKED EXAMPLE: BRINGING IT ALL TOGETHER

This section develops a worked example using the data tables held in the `prescribing.sqlite` database. This contains the full `prescriptions` data table, with some 120 million prescription records for 2018, from which the `prescribing_lite.sqlite` was sampled (all of the other data tables in the database are the same). This will take some time to download (it is 11.2 GB in size); you might wish to undertake another task, it will take a few hours to download.

```
download.file("http://archive.researchdata.leeds.ac.uk/734/1/
              prescribing.sqlite",
             "./prescribing.sqlite", mode = "wb")
```

As before, you should save it to your current working directory.

The task is to determine the number and cost per person of opioid prescribing in each LSOA. The aim is to illustrate how the sequences of operations can be combined to achieve an analytical goal, using a real-world example and a very large dataset. It is helpful to try to identify the steps in this. You may wish to have another look at Figure 4.1 to check that the data tables in the RSQLite database can be linked to support this query. The query needs to:

1. Extract the opioid data from the `prescriptions` table

2. Summarise the costs for each practice

3. Determine the proportion of patients in each LSOA from each practice

4. Determine the prescribing costs for each LSOA

5. Sum the costs of opioid prescriptions for each LSOA

6. Determine costs per person in the LSOA.

You should note that the same end point can be reached by undertaking the sequence of operations that constitute the query in a different order. However, it is instructive to think about the dplyr operations associated with each step in the query and how these relate to the different dplyr verbs and table joins:

1. 'Extract the opioid data' suggests the use of filter.

2. 'Summarise the costs for each practice' suggests using group_by and summarise.

3. 'Determine the proportion of patients in each LSOA from each practice' suggests using some kind of join function along with grouping and summarising operations.

4. 'Determine the prescribing costs for each LSOA' suggests using mutate.

5. 'Sum the costs of opioid prescriptions for each LSOA' suggests another mutate.

6. 'Determine costs per patient in the LSOA' suggests another join, summarise and mutate combination.

Finally, in this case we have a sampled subset of the full dataset in prescribing_lite.sqlite. This is convenient as it allows us to construct the query, run it over the data quickly and check the results of the intermediate steps (and that it is doing what we want) before it is applied it to the full database. You should reconnect to this now.

```
# connect to the database
db <- dbConnect(SQLite(), dbname="prescribing_lite.sqlite")
```

Step 1: Extract the opioid data

The bnf_code variable in the prescriptions table of the database contains values of the formal coding used in the UK for each prescription (see http://gmmmg.nhs.uk/html/formulary_bnf_chapters.html). This is hierarchical and codes individual drugs within families of drugs. Drugs starting with '040702' are opioid analgesics, for example (see https://ebmdatalab.net/prescribing-data-bnf-codes/).

The code below uses the %like% function to extract prescriptions that start with 040702, with no wildcard (%) at the front, but with one at the end of the pattern:

```
tbl(db, "prescriptions") %>%
  filter(bnf_code %like% '040702%')
```

Step 2: Summarise the opioid prescriptions costs

The code below extracts the opioid data from the prescriptions table as in step 1, and then summarises these over the GP practices using the group_by and summarise functions (remembering to ungroup each grouping):

```
tbl(db, "prescriptions") %>%
  filter(bnf_code %like% '040702%') %>%
  group_by(practice_id) %>%
  summarise(cost = sum(act_cost, na.rm = T)) %>%
  ungroup()
```

Step 3: Determine the proportion of patients in each LSOA from each practice

The patients data table has not been examined thus far. This is a long table of three fields with the counts of patients in each LSOA for each GP practice. The code below generates a table of the proportions of patients in each practice in each LSOA:

```
tbl(db, "patients") %>%
  group_by(practice_id) %>%
  summarise(prac_tot = sum(count, na.rm = T)) %>%
  ungroup() %>%
  left_join(tbl(db, "patients")) %>%
  mutate(prac_prop = as.numeric(count) / as.numeric(prac_tot))
```

What we would like to do is to combine the outputs of the code above with those of step 2 to be able to sum the total costs per LSOA. The code below does this with step 3 nested inside a left_join operation:

```
tbl(db, "prescriptions") %>%
  filter(bnf_code %like% '040702%') %>%
  group_by(practice_id) %>%
  summarise(cost = sum(act_cost, na.rm = T)) %>%
  ungroup() %>%
  left_join(
    tbl(db, "patients") %>%
      group_by(practice_id) %>%
      summarise(prac_tot = sum(count, na.rm = T)) %>%
      ungroup() %>%
      left_join(tbl(db, "patients")) %>%
      mutate(prac_prop = as.numeric(count) / as.numeric(prac_tot))
  )
```

Step 4: Determine the prescribing costs for each LSOA, for each practice

Here a simple `mutate` operation can be added to the code above to determine the costs associated with opioid prescribing for each LSOA, for each practice:

```
tbl(db, "prescriptions") %>%
  filter(bnf_code %like% '040702%') %>%
  group_by(practice_id) %>%
  summarise(cost = sum(act_cost, na.rm = T)) %>%
  ungroup() %>%
  left_join(
    tbl(db, "patients") %>%
      group_by(practice_id) %>%
      summarise(prac_tot = sum(count, na.rm = T)) %>%
      ungroup() %>%
      left_join(tbl(db, "patients")) %>%
      mutate(prac_prop = as.numeric(count) / as.numeric(prac_tot))
) %>%
mutate(lsoa_cost = cost*prac_prop)
```

Step 5: Sum the costs of opioid prescriptions for each LSOA

The `group_by` and `summarise` functions can be applied to sum the costs for each LSOA, removing any NA records. The code below also orders the table by total cost for illustration only:

```
tbl(db, "prescriptions") %>%
  filter(bnf_code %like% '040702%') %>%
  group_by(practice_id) %>%
  summarise(cost = sum(act_cost, na.rm = T)) %>%
  ungroup() %>%
  left_join(
    tbl(db, "patients") %>%
      group_by(practice_id) %>%
      summarise(prac_tot = sum(count, na.rm = T)) %>%
      ungroup() %>%
      left_join(tbl(db, "patients")) %>%
      mutate(prac_prop = as.numeric(count) / as.numeric(prac_tot))
) %>%
mutate(lsoa_cost = cost*prac_prop) %>%
group_by(lsoa_id) %>%
summarise(tot_cost = sum(lsoa_cost, na.rm = T)) %>%
ungroup() %>%
filter(!is.na(tot_cost)) %>%
arrange(desc(tot_cost))
```

Step 6: Determining the opioid prescription costs per person for each LSOA

To do this, we need the resident populations for each LSOA to determine rates per person. This is extracted from the social table, although a similar score could be derived from the patients data table:

```
tbl(db, "social") %>%
  mutate(population = as.numeric(population)) %>%
  select(lsoa_id, population)
      mutate(population = as.numeric(population)) %>%
      select(lsoa_id, population)
```

Again this can be added to the code block using a join with the step 4 code nested inside the join. A further mutate operation is then included to determine the costs per person and the results ordered again for illustration:

```
tbl(db, "prescriptions") %>%
  filter(bnf_code %like% '040702%') %>%
  group_by(practice_id) %>%
  summarise(cost = sum(act_cost, na.rm = T)) %>%
  ungroup() %>%
  left_join(
    tbl(db, "patients") %>%
      group_by(practice_id) %>%
      summarise(prac_tot = sum(count, na.rm = T)) %>%
      ungroup() %>%
      left_join(tbl(db, "patients")) %>%
      mutate(prac_prop = as.numeric(count) / as.numeric(prac_tot))
  ) %>%
  mutate(lsoa_cost = cost*prac_prop) %>%
  group_by(lsoa_id) %>%
  summarise(tot_cost = sum(lsoa_cost, na.rm = T)) %>%
  ungroup() %>%
  filter(!is.na(tot_cost)) %>%
  left_join(
    tbl(db, "social") %>%
      mutate(population = as.numeric(population)) %>%
      select(lsoa_id, population)
  ) %>%
  mutate(cost_pp = tot_cost/population) %>%
  arrange(desc(cost_pp))
```

Having developed the query using the prescribing_lite.sqlite database, this can now be closed and the connection to the full database opened:

```
# close the connection
dbDisconnect(db)
# connect to the full database
db <- dbConnect(SQLite(), dbname="prescribing.sqlite")
```

Then the code bock above can be run, replacing the last line (arrange(desc(cost_pp))) with collect() -> lsoa_result to pull the results of the query from the server and to assign these to a local in-memory R object. This will take a bit longer to run.

Finally, close the connection to the database:

```
dbDisconnect(db)
```

The results can be inspected and the summary indicates the presence of a record with NA values.

```
summary(lsoa_result)
##      lsoa_id                 tot_cost                population
##   Length:32935        Min.    :       2.74      Min.    : 983
##   Class :character    1st Qu. :    4464.48      1st Qu.: 1436
##   Mode  :character    Median  :    6238.25      Median : 1564
##                       Mean    :    6655.18      Mean    : 1614
##                       3rd Qu. :    8263.61      3rd Qu.: 1733
##                       Max.    :  130651.50      Max.    : 8300
##                                                 NA's    : 1
##         cost_pp
##   Min.    : 0.001494
##   1st Qu. : 2.829891
##   Median  : 4.001882
##   Mean    : 4.161286
##   3rd Qu. : 5.228562
##   Max.    :30.272204
##   NA's    :1

lsoa_result %>% arrange(desc(tot_cost))

## # A tibble: 32,935 x 4
##      lsoa_id    tot_cost population cost_pp
##      <chr>         <dbl>      <dbl>   <dbl>
##   1 NO2011      130651.         NA      NA
##   2 E01018076    63208.       2088    30.3
##   3 E01021360    34481.       2001    17.2
##   4 E01030468    32875.       3499     9.40
##   5 E01026121    32683.       2469    13.2
```

```
##  6 E01021770    31760.        2013    15.8
##  7 E01020797    30391.        2501    12.2
##  8 E01023155    29682.        2768    10.7
##  9 E01020864    28340.        2068    13.7
## 10 E01016076    27921.        2406    11.6
## # … with 32,925 more rows
```

On inspection, the NAs are unmatched LSOAs in the patients data table, likely for people in prisons or on military bases:

```
length(which(patients$lsoa_id == "NO2011"))
## [1] 4633
```

These can be removed:

```
lsoa_result %>% filter(!is.na(population)) -> lsoa_result
```

A quick examination and Google of the LSOAs with the highest cost and rates of prescribing rates per patient in England in 2018, E01018076 (Fenland in Cambridgeshire), E01021360 (Braintree in Essex) and E01025575 (Wyre near Blackpool) indicates that all are in the urban fringe:

```
lsoa_result %>% arrange(desc(cost_pp))
```

The E01018076 LSOA is a massive outlier (nearly seven times the median rate) and Figure 4.2 maps this location:

```
tmap_mode("view")
tm_shape(lsoa_sf[lsoa_sf$lsoa_id == "E01018076",]) +
    tm_borders(lwd = 2) +
        tm_basemap(server= "OpenStreetMap")
tmap_mode("plot")
```

A more informative national map is shown in Figure 4.3 (this will take some time to render!). This clearly shows higher opioid prescribing rates in the rural fringe, coastal areas and in the post-industrial areas of the North:

```
lsoa_sf %>% left_join(lsoa_result) %>%
    tm_shape() +
    tm_fill("cost_pp", style = "quantile", palette = "GnBu",
            title = "Rate per person", format = "Europe_wide")+
    tm_layout(legend.position = c("left", "top"))
```

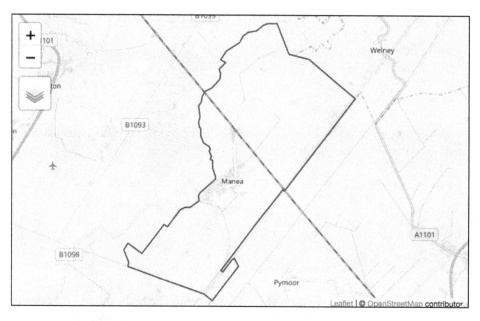

Figure 4.2   The LSOA with the highest opioid prescribing rates per person in 2018 with an OSM backdrop (© OpenStreetMap contributors)

4.6 SUMMARY

This chapter has illustrated how database queries can be constructed using the dplyr verbs and joins to wrangle data held in different database tables. These were used to construct a complex query that integrated and pulled data from different tables in the on-file database, but which required some of the dplyr functions to be replaced with workarounds, particularly for pattern matching. A key point is that dplyr tries to be as *lazy* as possible by never pulling data into R unless explicitly requested. In effect all of the dplyr commands are compiled and translated into SQL before they are passed to the database in one step. In this schema, the results of queries (e.g. tbl) create references to the data in the database and the results that are returned and printed to the console are just summaries of the query – the data remain on the database. The dplyr query results can be returned using collect, which retrieves data to a local tibble. The SQL created by dplyr can be examined with the show_query() function.

   The examples here used in-memory and local databases. Of course they can also be connected to remotely and the dplyr team provide a hypothetical example of the syntax for doing that (see https://db.rstudio.com/dplyr/):

```
con <- DBI::dbConnect(RMySQL::MySQL(),
    host = "database.rstudio.com",
    user = "hadley",
    password = rstudioapi::askForPassword("Database password")
)
```

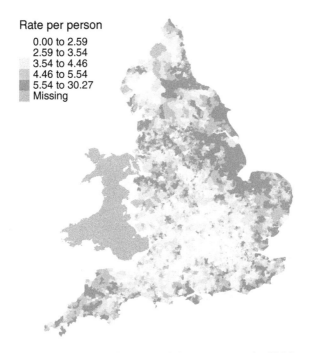

Rate per person
0.00 to 2.59
2.59 to 3.54
3.54 to 4.46
4.46 to 5.54
5.54 to 30.27
Missing

Figure 4.3   The spatial variation in opioid prescribing per person in 2018

dplyr and related packages such as stringr, many of which are loaded with the tidyverse package, contain many functions for manipulating and linking data tables. These wrap SQL and other functions such as pattern matching into a standard format to support data science. However, there are still some gaps in what they can do. In such cases, a bit of creative thought can usually overcome the problem.

You should save the query output as we will be using this for further analysis in later chapters:

```
save(list = c("lsoa_result", "lsoa_sf", "social"),
     file = "lsoa_result.RData")
```

REFERENCES

Horton, N. J., Baumer, B. S. and Wickham, H. (2015) Setting the stage for data science: Integration of data management skills in introductory and second courses in statistics. Preprint, arXiv: 1502.00318.

Rowlingson, B., Lawson, E., Taylor, B. and Diggle, P. J. (2013) Mapping English GP prescribing data: A tool for monitoring health-service inequalities. *BMJ Open*, 3(1), e001363.

5

EDA AND FINDING
STRUCTURE IN DATA

5.1 OVERVIEW

The preceding chapters have introduced data and spatial data (Chapter 2), the tools in the `dplyr` package and their use in manipulating and linking data and, critically, within chains of operations in a piping syntax (Chapter 3), and how databases for very large datasets can be created and held on file locally but outside of R's internal memory and then queried using piped `dplyr` functions (Chapter 4). As yet, little *formal* consideration has been given to exploratory data analysis (EDA), although some has been implicitly undertaken in the form of scatter plots and data summaries (as well as lollipop plots in Chapter 2).

The aim of EDA is to generate understanding of the data by revealing patterns, trends and relationships within and among the data variables. EDA generates summaries of data properties (data distribution, central tendencies, spread, etc.) and correlations with other variables, and reports these using tables or graphics. The first part of this chapter illustrates the core techniques in EDA.

However, standard EDA in this way provides only limited information about the *structure* of the data and the multivariate interactions and relationships between sets of variables the data contain. More nuanced and multivariate data understanding is needed to support hypothesis development and testing and inference (such as are undertaken in Chapter 6). Key to this are approaches for examining data and spatial data structure.

The following packages will be required for this chapter:

```
library(tidyverse)
library(RColorBrewer)
library(GGally)
library(data.table)
library(sf)
library(ggspatial)
library(tmap)
```

```
library(grid)
library(gridExtra)
```

The data used in this chapter combine the results of the opioid prescribing analysis that was written to lsoa_result.RData and other data tables in ch4_db.RData from Chapter 4. You should make sure these are in your current working directory and load them:

```
load("lsoa_result.RData")
load("ch4_db.RData")
ls()
```

5.2 EXPLORATORY DATA ANALYSIS

EDA is a critical part of any data analysis. It underpins the choice and development of methods and approaches of analysis, including which variables to retain and, sometimes, which scale(s) to operate at. This section covers the following:

- An introduction to the ggplot2 package
- EDA of individual variables
- Multivariate EDA
- Writing graphics, plots and figures to files (e.g. PNG, PDF, TIFF).

The aim of EDA is to understand data properties and data structure. Such activities support the development of analysis in the following typical situations:

1. Hypothesis development – you are searching through the haystack of data looking for the needle of opportunity and you want to confirm the approach you are taking.

2. Analysis – you are undertaking your analysis to test your hypothesis and you want confirmation that the analysis is heading in the right direction.

3. Communication – you have moved through the data and the problem, identified the problem or trend you are interested in and want to share the results with others.

Through EDA, it is possible to explore data properties (distributions, data spread, central tendencies, etc.) and dataset structure (how different variables interact with each other), and a number of visual and numeric tools are available.

R contains many functions for examining numeric data properties and some of these have been introduced in earlier chapters:

```
summary(lsoa_result)      # summary
mean(social$unemployed)   # mean
sd(social$unemployed)     # standard deviation
var(social$unemployed)    # variance
IQR(social$unemployed)    # interquartile range
```

These can be called for individually named variables as above. However, they can also be piped, usually along with some other operation, such as grouping. The code below summarises the mean and standard deviation of unemployment proportion overall and then does the same for different Output Area Classification (OAC) classes, the geo-demographic classification introduced in Chapter 4:

```
# overall
social %>%
   summarise(meanUn = mean(unemployed), sdUN = sd(unemployed))
# grouped by OAC
social %>%
   group_by(oac) %>%
   summarise(Count = n(), Mean_Un = mean(unemployed), SD_Un =
sd(unemployed)) %>%
   ungroup() %>%
   arrange(desc(Mean_Un))
```

It is also possible to examine correlations:

```
round(cor(social[, c(4, 5:11)]), 3)
##               unemployed noqual 14qual ptlt15 pt1630 ft3148
## unemployed         1.000  0.670 -0.536 -0.481 -0.064 -0.426
## noqual             0.670  1.000 -0.843 -0.443  0.201 -0.387
## 14qual            -0.536 -0.843  1.000  0.284 -0.317  0.259
## ptlt15            -0.481 -0.443  0.284  1.000  0.100 -0.201
## pt1630            -0.064  0.201 -0.317  0.100  1.000  0.000
## ft3148            -0.426 -0.387  0.259 -0.201  0.000  1.000
## ft49              -0.570 -0.621  0.707  0.218 -0.279  0.218
## llti               0.396  0.791 -0.595 -0.331  0.137 -0.402
##                  ft49   llti
## unemployed     -0.570  0.396
## noqual         -0.621  0.791
## 14qual          0.707 -0.595
## ptlt15          0.218 -0.331
## pt1630         -0.279  0.137
```

```
## ft3148        0.218 -0.402
## ft49          1.000 -0.478
## llti         -0.478  1.000
```

The significance of these correlations can be tested (note that Pearson's product-moment correlation coefficient is the default):

```
cor.test(~unemployed+noqual, data = social)
```

This shows that the correlation is highly significant (very low p-value) and also gives a 95% confidence interval of the correlation. An alternative format is as follows:

```
cor.test(social$unemployed, social$noqual)
```

Key Points

- EDA aims to provide understandings of data properties (distributions, data spread, central tendencies, etc.) and dataset structure (how different variables interact with each other).

- EDA supports hypothesis development and the choice and development of methods.

- It is critical for understanding and communicating the results of analysis.

- Data visualisations build on standard numeric approaches for exploring data properties and structure.

5.3 EDA WITH ggplot2

Visual approaches are very useful for examining single variables and for examining the interactions of variables together. Common visualisations include histograms, frequency curves and bar charts, and scatter plots, some which have been already introduced in this and earlier sessions. The ggplot2 package (Wickham, 2016) supports many types of visual summary. It allows multiple, simultaneous visualisations, and supports grouping by colour, shape and facets. The precise choice of these, out of the many possible options, requires some careful thinking. Do the data need to be sorted or faceted? Should transparency be used? Do I need to group the data? And so on.

The ggplot2 package is installed with the tidyverse package. R has several systems for making visual outputs such as graphs, but ggplot2 is one of the most elegant and most versatile. It implements the *grammar of graphics* (Wilkinson, 2012), a coherent system for describing and constructing graphs using a *layered* approach. If you would like to learn more about the theoretical underpinnings of ggplot2, Wickham (2010) is recommended.

The remainder of this chapter provides a rounded introduction to ggplot2, but note that it is impossible to describe *all* of the different parameters and visualisation options that are available. For this reason, the aims here are to establish core skills in visualising data with the ggplot2 package.

5.3.1 ggplot basics

The ggplot2 package provides a coherent system for describing and building graphs, implementing the grammar of graphics. The basic idea is that graphs are composed of different layers each of which can be controlled. The basic syntax is:

```
# specify ggplot with some mapping aesthetics
ggplot(data = <data>, mapping = <aes>) +
   geom_<function>()
```

In this syntax first ggplot is called, a dataset is specified (<data> in the above) and some mapping aesthetics are defined (<aes>). Together, these tell ggplot which variables are to be plotted (e.g. on the *x*- and *y*-axes), which variables are to be grouped and coloured. Finally, geom_<function>() specifies the plot type, such as geom_point(), geom_line(). ggplot has a default set of plot and background colours, line types and sizes for points, line width, text and so on, all of which can be overwritten.

These basics are probably best described through illustration. The code below generates a scatter plot of LSOA unemployment proportions against the proportion of the population lacking qualifications (noqual) from the social data, with a transparency term (alpha). The scatter plot is assigned to p:

```
p = ggplot(data = social, aes(x = unemployed, y = noqual)) +
   geom_point(alpha = 0.1)
```

p can be called to generate the plot:

```
p
```

Alternatively, further graphical elements can be added as layers. The example below fits a linear trend line:

```
p + geom_smooth(method = "lm")
```

This is equivalent to:

```
p = ggplot(data = social, aes(x = unemployed, y = noqual)) +
   geom_point(alpha = 0.1) +
   geom_smooth(method = "lm")
p
```

The ordering and location of information passed to the different elements in the plot are important. The code below will return an error because the call to ggplot() specifies the data but does not set any global mapping parameters or *aesthetics*. The result is that geom_smooth() does not know what to plot:

```
ggplot(social) +
    geom_point(aes(x = unemployed, y = noqual), alpha = 0.1) +
    geom_smooth(method = "lm")
```

This could be corrected by locally setting the mapping parameters within geom_smooth():

```
ggplot(social) +
    geom_point(aes(x = unemployed, y = noqual), alpha = 0.1) +
    geom_smooth(aes(x = unemployed, y = noqual), method = "lm")
```

However, it is more efficient to do this globally as was originally done:

```
ggplot(data = social, aes(x = unemployed, y = noqual)) +
    geom_point(alpha = 0.1) +
    geom_smooth(method = "lm")
```

This hints at how control can be exercised over different plot layers, specifying different data, different aesthetics/parameters, and so on.

Layer-specific parameters can be set such as colour, shape, transparency, size, thickness, and so on, depending on the graphic, to change the default style. Additionally, the ggplot2 package includes a number of predefined styles, called by theme_X(), that can be added as a layer. The code below applies theme_minimal() as in Figure 5.1:

```
ggplot(data = social, aes(x = unemployed, y = noqual)) +
    # specify point characteristics
    geom_point(alpha = 0.1, size = 0.7, colour = "#FB6A4A", shape = 1) +
    # specify a trend line and a theme/style
    geom_smooth(method = "lm", colour = "#DE2D26") +
    theme_minimal()
```

You should explore other ggplot themes, which are listed if you enter ?theme_bw at the console.

5.3.2 Groups with ggplot

An important consideration is the different ways that ggplot can be used to display groups of data. Groups can be defined in a number of different ways

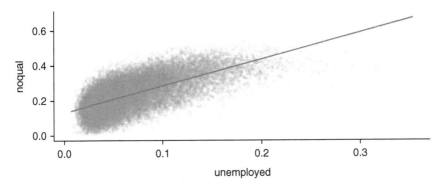

Figure 5.1 A scatter plot of lack of qualifications against rates of unemployment

with ggplot. These might be different treatments, classes or other discrete categories within the data, or groups that are defined either independently or as part of a piped operation. There are two basic plotting options for displaying groups.

First, multiple groups can be displayed in the same plot if they are specified within the mapping aesthetics, using parameters such as size, colour, shape. The code below does this with colour as the main grouping parameter. Notice how, having been grouped, the geom_smooth function plots a trend line for each group:

```
ggplot(data = social, mapping = aes(x = unemployed,
                                    y = noqual, colour = oac)) +
    geom_point(size = 0.05, alpha = 0.3) +
    scale_colour_brewer(palette = "Set1") +
    # specify a trend line
    geom_smooth(method = "lm")
```

The second option is to generate individual plots for each group using some kind of faceting. The code below generates separate scatter plots for each OAC type as in Figure 5.2:

```
ggplot(data = social, aes(x = unemployed, y = noqual)) +
    # specify point characteristics
    geom_point(alpha = 0.1, size = 0.7, colour = "#FB6A4A", shape = 1) +
    # add a trend line
    geom_smooth(method = "lm", colour = "#DE2D26") +
    # specify the faceting and a theme/style
    facet_wrap("oac", nrow = 2) +
    theme_minimal()
```

There are many further examples of facets through this chapter.

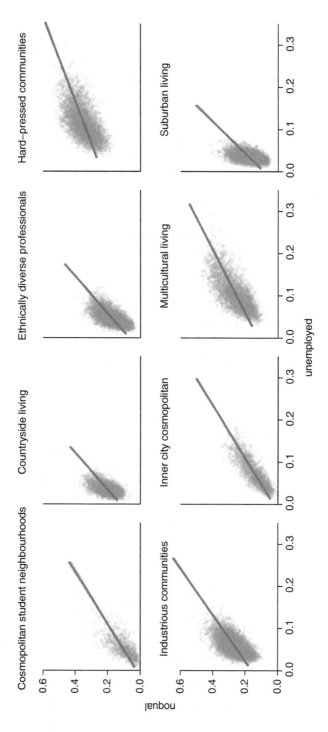

Figure 5.2   A faceted scatter plot of rates of unemployment against lack of qualifications

It should be evident that it is possible to control every element in the graphic with ggplot2. However, it is beyond the scope of this chapter to describe or illustrate every possible option for exercising control, which is why ggplot2 has books dedicated to it – see the references listed at the end of the chapter. As you work through these examples and play with the parameters, you should start to gather your own library of techniques and approaches for controlling plots in different ways.

Key Points

- The ggplot2 package supports many types of data visualisation.

- It uses a *layered* approach to construct graphics, and each individual layer can be controlled.

- Each use of ggplot requires mapping *aesthetics* to be specified. These describe which variables are to be plotted and grouped.

- There are default plot types, colours and styles (themes), all of which can be overwritten.

- In the layered approach, the ordering and location of layer information passed to the different elements in the plot are important.

- It is possible to visualise separate groups in the data by specifying a grouping aesthetic (using parameters such as size, colour, shape), or through faceting to produce a series of individual plots.

5.4 EDA OF SINGLE CONTINUOUS VARIABLES

Starting with the simplest case, the distribution of a single continuous variable can be visualised using a density plot, a histogram or a boxplot. Essentially the density plot is a smoothed version of the histogram. These provide similar information in a visual format to that returned by the summary() and fivenum() functions. The code below does this for the limiting long-term illness variable (llti) in the social dataset, while the bins parameter controls the number of histogram groups (columns):

```
# a density plot
ggplot(social, aes(x = llti)) +
  geom_density()
# a histogram
ggplot(social, aes(x = llti)) +
  geom_histogram(bins = 30, col = "red", fill = "salmon")
```

```
# a boxplot
ggplot(social, aes(x = "", y = llti)) +
  geom_boxplot(fill = "dodgerblue", width = 0.2) +
  xlab("LLTI") + ylab("Value")
```

A *density histogram* can also be constructed by specifying a density aesthetic for y in the geom_histogram function (y=..density..). The areas of each bin indicate the relative probability of the bins in the distribution (the sum of the bin counts is 1). A density function can also be added to the histogram as in Figure 5.3 using the code below:

```
ggplot(social, aes(x = llti)) +
  geom_histogram(aes(y=..density..),bins = 30,
                 fill="indianred", col="white") +
  geom_density(alpha=.5, fill="#FF6666")
```

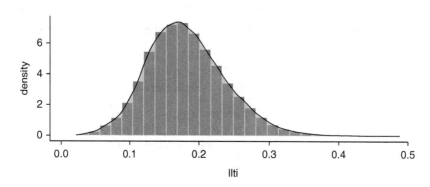

Figure 5.3   A density histogram with a probability density function

The line geom_histogram(aes(y=..density..)...) rescales the histogram counts so that bar areas integrate to 1, but is otherwise the same as the standard ggplot2 histogram. The geom_density() function adds a density curve to the plot with a transparency term (alpha=.5). You could plot just the density curve by running the code below:

```
ggplot(social, aes(x = llti)) +
  geom_density(aes(y=..density..), alpha=.5, fill="#FF6666")
```

Recall that each use of ggplot requires a set of aesthetic mappings to be used and that these are specified using the aes argument. They can be globally specified when ggplot is used to *initialise* the plot or when the particular layer of the plot is called. The code below generates the same histogram plot as the above, but this time with all of the mapping aesthetics specified for each plot layer individually:

```
ggplot() +
  geom_histogram(data = social, aes(x = llti, y=..density..),
                  bins = 30, fill="indianred", col="white") +
  geom_density(data = social, aes(x = llti),
              alpha=.5, fill="#FF6666")
```

This *layered* approach to ggplot operations, and in the use of density functions, allows different variables to be passed to different plot layers. The code below shows llti along with unemployed. Notice how the y parameter in the initial call to ggplot aesthetics is used by both geom_histogram and geom_density:

```
ggplot(data = social, aes(y=..density..)) +
  geom_histogram(aes(x = llti), bins = 30,
                  fill="indianred", col="white") +
  geom_density(aes(x = noqual), alpha=.5, fill="#FF6666") +
  xlab("LLTI (bins) with lack of qualifications (curve)")
```

The density plots can be further extended to compare the density plots of different groups, showing how different levels of llti are associated with different OAC classes:

```
ggplot(social, aes(llti, stat(count), fill = oac)) +
  geom_density(position = "fill") +
  scale_colour_brewer(palette = "Set1")
```

The stat(count) argument returns the density by number of points. Other options include density, scaled and ndensity. Notice the use of scale_colour_brewer and Set1. The latter is one of a number of different colour scales with the specific naming relating to their use. The RColorBrewer palettes can be listed:

```
library(RColorBrewer)
display.brewer.all()
brewer.pal(11, "Spectral")
brewer.pal(9, "Reds")
```

It is also possible to combine a series of individual histograms using the facet_wrap function as in Figure 5.4:

```
ggplot(social, aes(x = llti)) +
  geom_histogram(fill="firebrick3", bins = 30, col = "white")+
  facet_wrap( ~ oac, nrow = 2, scales = "fixed")
```

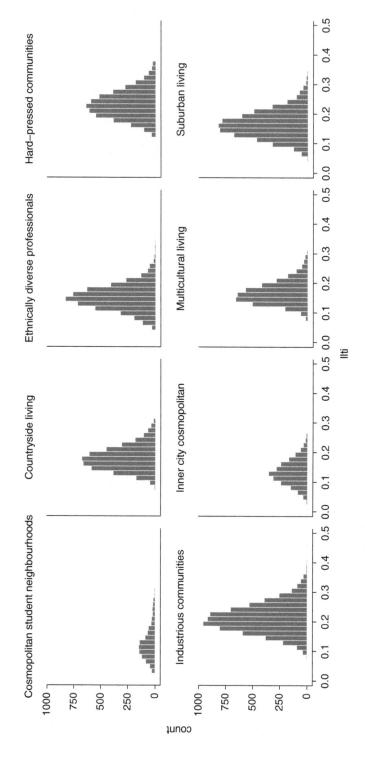

Figure 5.4 Histograms of different `llti` rates in the OAC classes

Notice the parameters that are passed to `facet_wrap()`. They define the variable that is to be *faceted*, the number of rows in the plot columns and whether the scales for the plots are to be fixed or not (the default is `fixed` for comparison). The first argument of `facet_wrap()` should be a formula, which you create with ~ followed by a variable name (here 'formula' is the name of a data structure in R, not a synonym for 'equation'). The variable that you pass to `facet_wrap()` should be discrete (i.e. categorical).

The analysis can be modified for grouped boxplots. The example removes the legend and gets rid of the default `ggplot2` colour scheme:

```
social %>%
  ggplot(aes(y=llti,fill = oac)) +
    # specify the facets
    facet_wrap(~oac, ncol = 2) +
    # flip the coordinates and specify the boxplot
    coord_flip() + geom_boxplot() +
    # specify no legend and the colour palette
    theme(legend.position = "none") +
    scale_fill_manual(name = "OAC class",
                      values = brewer.pal(8, "Spectral"))
```

Notice how the piping syntax was used with `ggplot` in the above, effectively piping the first argument (`data`).

These examples illustrate how additional variables, particularly categorical variables, can be used to split the plot into *facets*, or subplots that each display one subset of the data.

This can be further refined by *ordering* the OAC classes by their median `llti` values and plotting them, using a number of other plot parameters as in Figure 5.5:

```
social %>%
  ggplot(aes(x = reorder(oac, llti, FUN = median),
             y=llti, fill = oac)) +
    # specify the boxplot
    geom_boxplot(aes(group = oac), outlier.alpha=0.4,
                 outlier.colour="grey25") +
    # specify the colour palette
    scale_fill_manual(name = "OAC class",
                      values = brewer.pal(8, "Spectral")) +
    # flip the coordinates and specify the axis labels
    coord_flip() + xlab("") + ylab("LLTI") +
    # specify some styling
    theme_minimal() + theme(legend.position = "none")
```

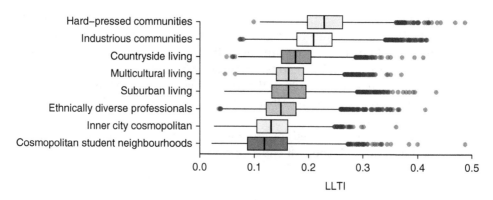

Figure 5.5 Boxplots of `llti` against OAC classes, ordered by the class median value

There is quite a lot going on in the code above. It is instructive to unpick some of it. You should note (and play around with!) the specification of the `fill` aesthetics that tells `ggplot` which variables are to be shaded, the treatment of outliers, specifying their transparency and shading and the overriding of the default `ggplot2` palette with `scale_fill_manual()`. The `coord_flip()` function transposes the *x*- and *y*-axes and their legends. Finally, one of the `ggplot` *themes* was applied. This has a layout that includes a legend, meaning that a line of code to remove the legend has to be specified *after* the theme call if that is what is wanted.

Key Points

- Single, continuous variables can be visualised using density plots, histograms or boxplots.

- Density histograms indicate the relative probability of the bins, and because the sum of the bin counts is 1, they can be used to compare distributions of different continuous variables.

- Faceting and grouping can be used to compare across groups.

- Grouped boxplots can be ordered by some function (`mean`, `median`, etc.).

5.5 EDA OF MULTIPLE CONTINUOUS VARIABLES

Earlier scatter plots of individual pairs of variables were used to show correlations. The scatter plot provides the simplest way of examining how two variables interact. These are essentially visualisations of the correlations generated in Section 5.2. The code below plots the `ft49` and `llti` variables as an example of a simple scatter plot. `geom_smooth()` by default fits a trend line using a generalised additive model when there are more than 1000 observations and a loess model otherwise (in the code snippets above a linear trend line was fitted):

```
social %>% ggplot(mapping = aes(x = ft49, y = llti)) +
    geom_point(size = 0.2, alpha = 0.2) +
    geom_smooth()
```

It is possible to examine *all* of the pairwise relationships (correlations) between continuous variables in datasets using the default plot() function:

```
plot(social[, c(4, 5:11)], cex = 0.2, col = grey(0.145,alpha=0.2))
```

This will take some time run because of the size of the social dataset.

Another useful tool here is to show the upper triangle of the scatter plot matrix with smoothed trend lines. These are achieved with *loess* curve fits (Cleveland, 1979). In a similar way to the default geom_smooth methods, these are smooth bivariate trend lines and provide a good way of judging by eye whether there are useful correlations in the data, including collinearity in variables. Essentially a straight-line-shaped trend with not too much scattering of the individual points suggests collinearity might be an issue. When two predictors are correlated it can be difficult to identify whether it is one or the other (or both) that influence the quantity (response) to be predicted. The code below does this. Adding upper.panel=panel.smooth to the above code causes the loess curves to be added:

```
plot(social[, c(4, 5:11)], cex = 0.2, col = grey(0.145,alpha=0.2),
        upper.panel=panel.smooth)
```

A ggplot-like version of this is available using the ggpairs() function in the GGally package as in Figure 5.6. Note the use of sample_n function to randomly sample 2500 records from the data. If the plot does not display, make the plot window larger:

```
social %>% sample_n(2500) %>% select(-c(1,2, 12:15)) %>%
    ggpairs(lower = list(continuous = wrap("points",
                                            alpha = 0.3, size=0.1))) +
        theme(axis.line=element_blank(),
            axis.text=element_blank(),
            axis.ticks=element_blank())
```

This can be further enhanced by comparing different groups, here defined by splits around median unemployment. The code snippet below summarises the variables when they are split into high ("H") and low ("L") unemployment groups in this way. Also note the use of the dot (.) in place of the select function. This does the same thing but with a different syntax. The dot refers to the data that come out of the dplyr pipe. Here it is being used to select specific fields/columns. The plot in Figure 5.7 includes split distributions, scatter

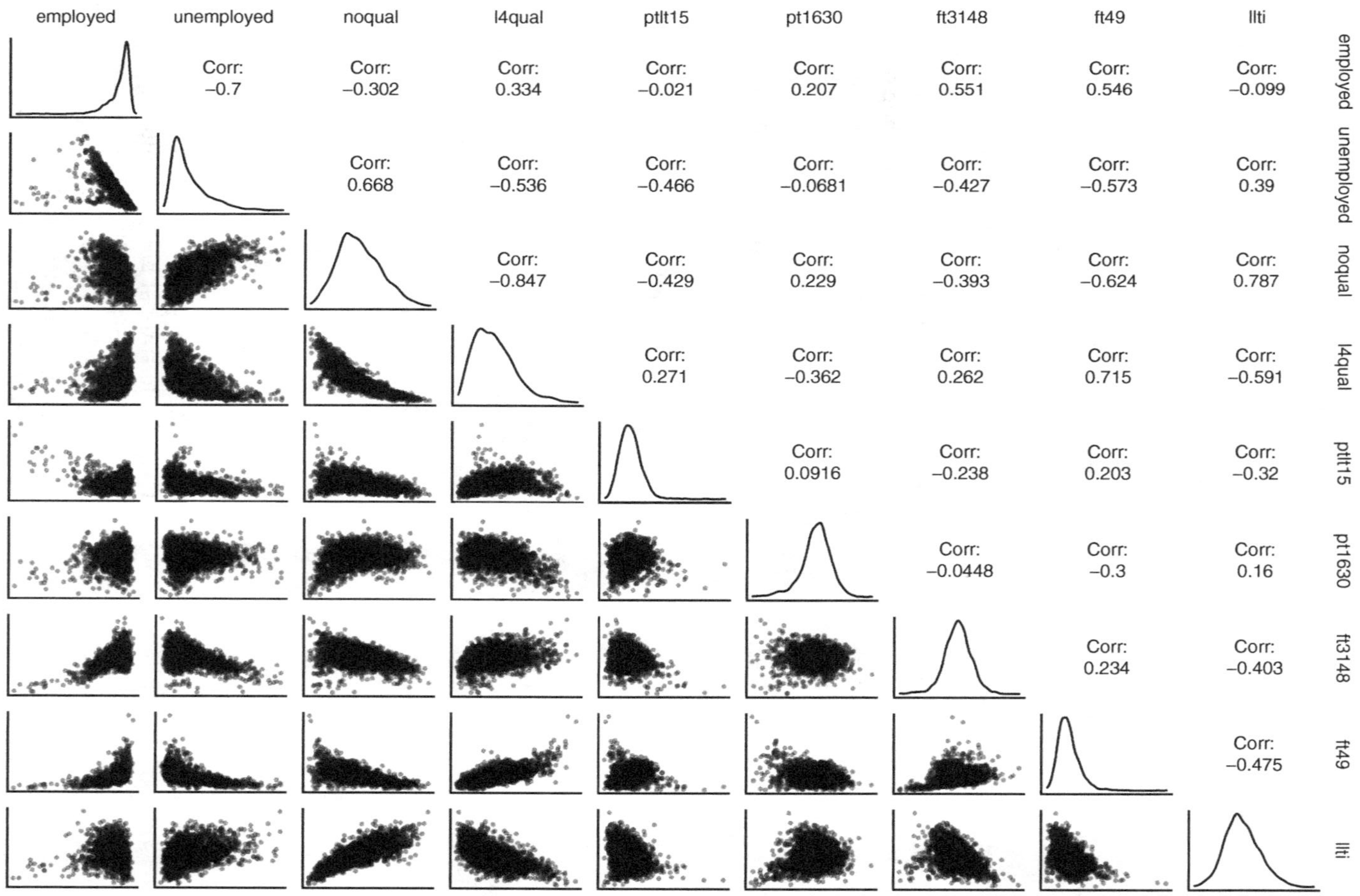

Figure 5.6 A `ggpairs` plot of a sample of the `social` data

plots and boxplots. A few additional parameters need to be passed to `ggpairs` to make it all work:

```
social %>% sample_n(2500) %>% .[-c(1,2, 12:15)] %>%
  mutate(unemp_high = ifelse(unemployed > median(unemployed),
    "H","L")) %>%
ggpairs(aes(colour = unemp_high, alpha = 0.4),
        upper = list(continuous = wrap('cor', size = 2.5)),
        lower = list(continuous =
                     wrap("points", alpha = 0.7, size=0.1))) +
theme(axis.line=element_blank(),
      axis.text=element_blank(),
      axis.ticks=element_blank())
```

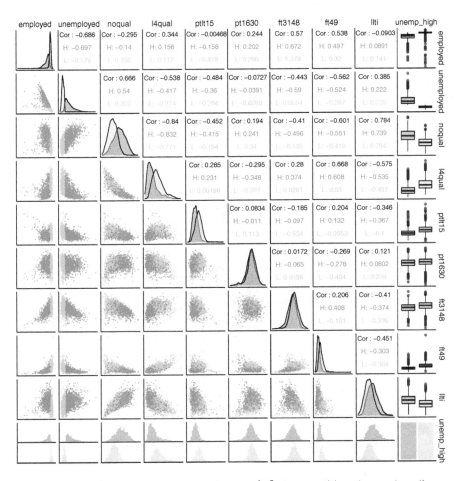

Figure 5.7 A `ggpairs` plot of a sample of the `social` data, partitioned around median unemployment

A different kind of visualisation of correlations is possible in ggplot using the geom_tile function along with the order.single() function from gclus, which sorts the data so that similar relationship pairs are adjacent. The code below does the following to generate Figure 5.8:

1. Extract the variables that are numeric.

2. Generate a correlation matrix.

3. Pivot the result into wide format using tools from the tidyr package.

4. Use the geom_tile function.

There is a lot going on in the code below, but the comments should help you navigate through it. Notice the use of scale_fill_gradient in the ggplot function. This is used to specify the shading and breaks that are passed to the item being *filled*, in this case the value variable in lengthened, specifying two colours from the RColorBrewer reds and blues palettes rather than the default ggplot treatment of red and blue. Also, geom_text prints the values of value to each tile, showing the actual correlation numerically.

```
# 1. create an index to extract the numeric variables
index = which(as.vector(unlist(sapply(social, class))) == "numeric")
df = social[, index]
# 2. construct the correlation matrix
cor.mat <- round(cor(df),3)
# 3. pivot the correlation matrix to long format
lengthened <- pivot_longer(as_tibble(cor.mat,rownames = "row"),-row)
# checks: uncomment the line below to see
# cor.mat
# head(lengthened)
# 4. then construct the main ggplot with geom_tile
ggplot(lengthened, aes(row, name, fill = value)) +
  geom_tile(color = "white") +
  scale_fill_gradient2(low = "#CB181D", high = "#2171B5",
                       mid = "white", midpoint = 0,
                       limit = c(-1,1), space = "Lab",
                       name="Correlation") +
  theme_minimal() +
  # make sure x and y have the same scaling
  coord_fixed() +
  # add the values to the tiles, using row and name as coordinates
  geom_text(aes(row, name, label = value),
            color = "black", size = 2) +
```

```
# specify theme elements
theme_minimal() +
# adjust text direction on x-axis
theme(axis.text.x = element_text(angle = 45,
                                 vjust = 1,hjust = 1)) +
# remove the axis labels by making them blank
xlab("") +ylab("")
```

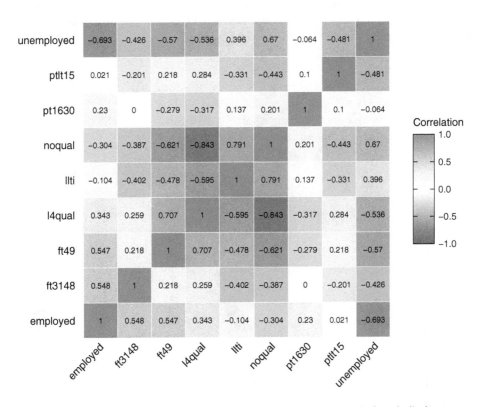

Figure 5.8 Correlation tile plot of numeric variables, ordered by correlation similarity

A further way of plotting correlations is to use hexbins (Figure 5.9). These are hexagonal tiles that are shaded to represent the number of data points at that location in a scatter plot. They provide a convenient way of summarising data. This is especially useful with very large data. Plot character size and transparency can be used as above to aid visualisation. There are a number of binning options: geom_bin2d, geom_hex and geom_density2d. You could try using each of these in turn in the second line of the code below to generate alternative versions of Figure 5.9. Note that you may have to install the hexbin package:

```
social %>%
    ggplot(mapping = aes(x = llti, y = unemployed)) +
        geom_hex() + labs(fill='Count') +
        scale_fill_gradient(low = "lightgoldenrod1", high = "black")
```

The bin shadings provide a convenient way of representing the density of data points with a similar value. Of course different elements of faceting, ordering, binning and grouping can be combined as in Figure 5.10:

```
social %>%
    ggplot(mapping = aes(x = llti, y = noqual)) +
        geom_hex(bins=15) +
        facet_wrap(~oac, nrow = 2) + labs(fill='Count') +
        scale_fill_gradient(low = "lemonchiffon", high = "darkblue")
```

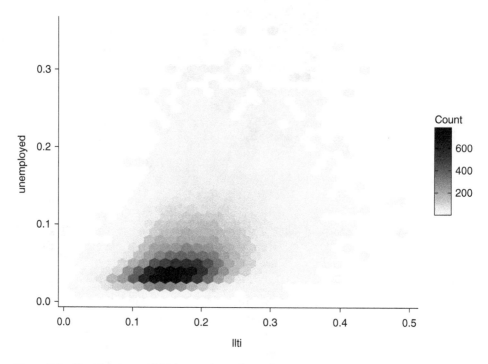

Figure 5.9 Hexbin plots of **llti** and **unemployed** variables

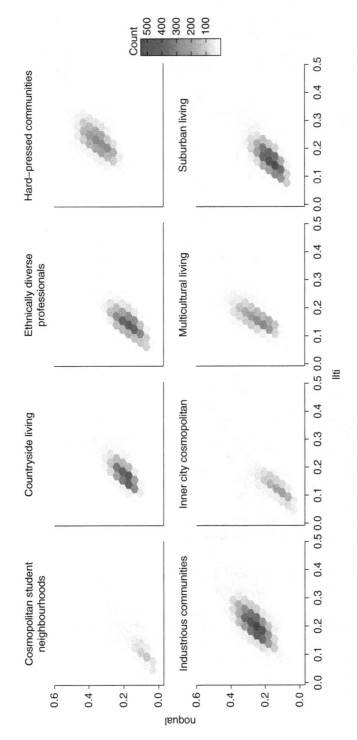

Figure 5.10 Hexbin plots of `llti` and `noqual` variables by OAC class

All the approaches until now have examined graphics showing pairwise relations. It is sometimes useful to examine multiple variables together. This can be done using *radar* plots using the `coord_polar()` function with an aggregate function to generate and display summaries (means in this case) of the variables:

```
social %>%
   # select the data
   select_if(is.numeric) %>%
   select(-oac_code, -population, -employed) %>%
   # generate variable means summaries
   mutate_if(is.numeric, mean) %>%
   # select a single row and transpose to something like
   # a tidy data frame with a fake "group"
   slice(1) %>% t() %>% data.frame(Group = "All") %>%
   # create a variable of the rownames and name the values
   rownames_to_column() %>%
   `colnames<-`(c("name", "value", "group")) %>%
   # arrange by name
   arrange(name) %>%
   # initiate the plot
   ggplot(aes(x=factor(name), y=value, group= group,
                colour=group, fill=group)) +
     # specify the points and areas for shading
     geom_point(size=2,) + geom_polygon(size = 1, alpha= 0.2) +
     # specify the polar plot
     coord_polar() +
     # apply some style settings
     theme_light() +
     theme(legend.position = "none",
            axis.title = element_blank(),
            axis.text = element_text(size = 6))
```

It also possible to visually explore the different general properties of different groups. The code below extends the radar plot above to compare across the OAC classes to generate Figure 5.11. Note the need to rescale the data, which then need to be made long using `pivot_longer` before being passed to `ggplot`. You may also wish to examine the intermediary outputs (the effects of the selection, data lengthening, etc.) by running blocks of code, remembering either to omit the last pipe or in this case to insert `head()` after each `%>%` (as the input is a `data.frame`).

```
social %>%
    # select the data
    select_if(is.numeric) %>% select(-oac_code) %>%
    # rescale using z-scores for the numeric variables
    mutate_if(is.numeric, scale) %>%
    # generate group summaries
    aggregate(by = list(social$oac), FUN=mean) %>%
    # lengthen to a tidy data frame
    pivot_longer(-Group.1) %>%
    # rename one of the OAC classes (for plot labels that fit!)
    mutate(Group.1 =
        str_replace(Group.1,
                    pattern = "Cosmopolitan student neighbourhoods",
                    replacement = "Cosmopolitan student areas")) %>%
    # sort by the variable names - this is needed for the plot
    arrange(name) %>%
    # initiate the plot
    ggplot(aes(x=factor(name), y=value,
               group= Group.1, colour=Group.1, fill=Group.1)) +
    # specify the points and areas for shading
    geom_point(size=2) + geom_polygon(size = 1, alpha= 0.2) +
    # specify the shading
    scale_color_manual(values= brewer.pal(8, "Set2")) +
    scale_fill_manual(values= brewer.pal(8, "Set2")) +
    # specify the faceting and the polar plots
    facet_wrap(~Group.1, ncol = 4) +
    coord_polar() +
    # apply some style settings
    theme_light() +
    theme(legend.position = "none", axis.title = element_blank(),
          axis.text = element_text(size = 6))
```

In the above code the data were rescaled to show the variables on the same polar axis. Rescaling can be done in a number of ways, and there are many functions to do it. Here the scale() function applies a classic standardised approach around a mean of 0 and a standard deviation of 1 (i.e. a z-score). Others use variable minimum and maximum to linearly scale between 0 and 1 (such as the rescale function in the scales package). The polar plots in the figure show that there are some large differences between classes in most of the variables, although how evident this is will depend on the method of rescaling and aggregation. Try changing scale to scales::rescale in the code above and changing mean to median to examine these.

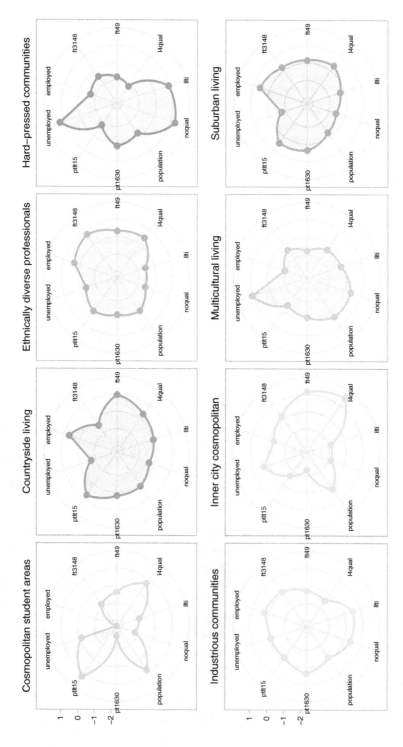

Figure 5.11 Radar plots of the variable mean values for each OAC class

Writing figures to file

Figures, plots, graphs, maps, and so on can be written out to a file for use in a document or report. There are a whole host of formats to which graphics can be written out, but usually we want PDFs, JPEGs, TIFFs or PNGs. The code below creates and writes a figure in PNG format. You will have to experiment with the different ways that the size and density (e.g. dots per inch, dpi) can be adjusted, which of course affect the plot display (text size etc.).

PDF, PNG and TIFF files can be written using the following functions:

```
pdf()
png()
tiff()
```

You should examine the help for these. The key thing you need to know is that these functions all open a file. The open file needs to have a map or the figure written to it and then needs to be closed using dev.off(). So the basic syntax is (do not run this code – it is just showing syntax):

```
pdf(file = "MyPlot.pdf", other settings)
<code to produce figure or map>
dev.off()
```

You should try to write out a .png file of the plot using the code below. This writes a .png file to the current working directory, which can always be changed:

```
# open the file
png(filename = "Figure1.png", w = 7, h = 5, units = "in", res = 300)
# make the figure
social %>% ggplot(mapping = aes(x = llti, y = unemployed)) +
    geom_hex() + labs(fill='Count') +
    scale_fill_gradient(low = "lightgoldenrod1", high = "black")
# close the file
dev.off()
```

Key Points

- Scatter plots of individual pairs of variables are the core technique for visualising their correlations.

- Adding trend lines of different forms can help to show these relationships.

- The GGalley package provides a number of very powerful tools for visualisations that are built on ggplot2.

- Multiple correlations (as in a correlation matrix) can be visualised using `geom_tile()`.

- Pairwise correlations for very large numbers of observations can be visualised using different summary bins and contours.

- Multivariate data properties can be visualised using radar plots, and these can be used to compare the multivariate properties of different groups.

5.6 EDA OF CATEGORICAL VARIABLES

Categorical data typically describe the *class* of an observation. This section describes some basic approaches for examining categorical variables and describes techniques for visualising them and how they interact.

5.6.1 EDA of single categorical variables

Individual categorical variables can be compared by examining their frequencies. Frequencies of a single categorical variable can be examined numerically with tables and visually with some kind of bar or column plot. The code below calculates the frequencies numerically for the counts of OAC classes in the `social` data table in different ways using base R and `dplyr` functionality:

```
# using table in base R
as.data.frame(table(social$oac))
# using count from dplyr
count(social, oac)
# ordered by count/n
arrange(count(social, oac), n)
# in descending order
arrange(count(social, oac),-n)
# piped
social %>% count(oac) %>% arrange(-n)
```

Such frequency counts can be extended to include proportions by combining with a mutate operation:

```
social %>% count(oac) %>% arrange(-n) %>% mutate(freq = n / sum(n))
## # A tibble: 8 x 3
##     oac                               n     freq
##     <fct>                          <int>   <dbl>
## 1 Industrious communities          7169   0.206
```

```
## 2 Suburban living                           6023   0.173
## 3 Ethnically diverse professionals          5206   0.150
## 4 Hard-pressed communities                  4828   0.139
## 5 Countryside living                        4381   0.126
## 6 Multicultural living                      3838   0.110
## 7 Inner city cosmopolitan                   2170   0.0624
## 8 Cosmopolitan student neighbourhoods       1138   0.0327
```

The class frequencies can also be plotted using the geom_bar function with ggplot. The code snippets below start with a basic plot and build to more sophisticated visualisations:

```
# standard plot
ggplot(social, aes(x = factor(oac_code))) +
  geom_bar(stat = "count") + xlab("OAC")
# ordered using fct_infreq() from the forcats package
# (loaded with tidyverse)
ggplot(social, aes(fct_infreq(factor(oac_code)))) +
  geom_bar(stat = "count") + xlab("OAC")
# orientated a different way
ggplot(social, aes(y = factor(oac))) +
  geom_bar(stat = "count") + ylab("OAC")
# with colours - note the use of the fill in the mapping aesthetics
ggplot(social, aes(x = factor(oac_code), fill = oac)) +
  geom_bar(stat = "count")
# with bespoke colours with scale_fill_manual and a brewer palette
ggplot(social, aes(x = factor(oac_code), fill = oac)) +
  geom_bar(stat = "count") +
  scale_fill_manual("OAC class label",
                    values = brewer.pal(8, "Set1")) +
  xlab("OAC")
```

We can extend the bar plot to include the counts of specific components. The code below uses a piped operation to calculate grouped summaries of the total number of people with no qualifications (NoQual) and with level 4 qualifications (L4Qual) in each OAC class, which are then passed to different plot functions to create Figure 5.12. (Note that these are similar to the plots created by the code above but not exactly the same due to the different populations in each LSOA.) The code snippet below uses different approaches to show the results of the grouping, with different parameters used in the ggplot aesthetics and in the geom_bar parameters. Each plot is assigned to an R object and then these are combined using the grid.arrange function in the gridExtra package to generate Figure 5.12.

```
# using a y aesthetic and stat = "identity" to represent the values
g1 = social %>% group_by(oac_code) %>%
  # summarise
  summarise(NoQual = sum(population * noqual),
            L4Qual = sum(population * l4qual)) %>%
  # make the result long
  pivot_longer(-oac_code) %>%
  # plot
  ggplot(aes(x=factor(oac_code), y=value, fill=name)) +
  geom_bar(stat="identity", position=position_dodge()) +
  scale_fill_manual("Qualifications",
                    values = c("L4Qual"="red","NoQual"="orange")) +
  xlab("OAC") + theme_minimal()
# using just an x aesthetic and stat = "count" with fill and weight
g2 = social %>% group_by(oac_code) %>%
  # summarise
  summarise(NoQual = sum(population * noqual),
            L4Qual = sum(population * l4qual)) %>%
  # make the result long
  pivot_longer(-oac_code) %>%
  # plot
  ggplot(aes(factor(oac_code))) +
  geom_bar(stat="count", aes(fill=name, weight=value)) +
   scale_fill_manual("Qualifications",
                    values = c("L4Qual"="red","NoQual"="pink")) +
  xlab("OAC") + theme_minimal()
# plot both plots
gridExtra::grid.arrange(g1, g2, ncol=2)
```

A different approach would be to use geom_col() instead of geom_bar() in a hybrid of the plots creating g1 and g2 above. The geom_col() function requires mapping aesthetics for both x and y to be specified (as in g1) and then can be specified with or without position=position_dodge() to create either g1 (with) or g2 (without), to have the bars stacked side by side or on top of each other, respectively. Try running the code below first with geom_col() commented out and then uncommented:

```
social %>% group_by(oac_code) %>%
  summarise(NoQual = sum(population * noqual),
            L4Qual = sum(population * l4qual)) %>%
  pivot_longer(-oac_code) %>%
  ggplot(aes(x=factor(oac_code), y=value, fill=name)) +
  # Option 1: stacked on top
  geom_col() +
```

```
# Option 2: stacked side by side (uncomment the line below)
# geom_col(position=position_dodge()) +
scale_fill_manual("Qualifications",
                  values = c("L4Qual"="red","NoQual"="pink")) +
xlab("OAC") + theme_minimal()
```

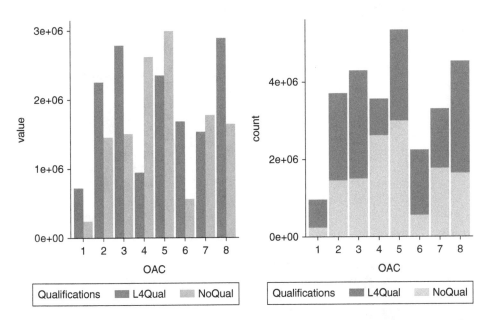

Figure 5.12   Bar plots of OAC class with the proportions of people with different levels of qualifications, created and displayed in different ways

5.6.2 EDA of multiple categorical variables

Previous sections contained many examples of visualisations of different categorical variables in terms of their continuous variable properties. This was done using grouped or faceted plots of single variables (boxplots in Figure 5.6), individual pairwise correlations (scatter plots in Figure 5.2), multiple pairwise scatter plots (Figures 5.6, 5.7 and 5.9) or radar plots of multiple properties (Figure 5.11). All of these sought to show the differences in numeric variables between groups, in this case OAC classes. Implicitly these have shown the relationships between continuous variables by categorical groups.

However, it is also possible to examine how different categorical variables interact using a correspondence table. This is commonly done to assess accuracy in categorical mapping such as land cover and land use. The table is used to compare observed classes (e.g. collected by field survey) with those predicted by a remote sensing classification. Here it can be used to compare the correspondences between different categories in the social dataset which contains a Rural–Urban Classification (RUC) as well as the OAC. The RUC contains eight classes for the LSOA areas for England and Wales. Classes A1, B1, C1 and C3 are urban classes and D1, D2, E1 and E2 are rural classes.

The relationship between the RUC classes and the OAC classes can be simply summarised using different functions for tabulating data counts in a correspondence table:

```
social %>%
  select(oac, ruc11_code) %>% xtabs( ~ ., data = .)
##                                           ruc11_code
## oac                                        A1    B1    C1    C2
##    Cosmopolitan student neighbourhoods    271    73   784     6
##    Countryside living                      15     8   536    15
##    Ethnically diverse professionals      2158   123  2760     1
##    Hard-pressed communities              1706   281  2727    19
##    Industrious communities               1142   329  4252    50
##    Inner city cosmopolitan               2121     3    46     0
##    Multicultural living                  2724   131   983     0
##    Suburban living                       1386   260  3636     3
##                                           ruc11_code
## oac                                        D1    D2    E1    E2
##    Cosmopolitan student neighbourhoods      1     1     2     0
##    Countryside living                    1082    80  2319   326
##    Ethnically diverse professionals       129     3    32     0
##    Hard-pressed communities                88     7     0     0
##    Industrious communities               1228   105    61     2
##    Inner city cosmopolitan                  0     0     0     0
##    Multicultural living                     0     0     0     0
##    Suburban living                        661     1    76     0
```

Notice the parameters that are passed to the xtabs function. xtabs is not a piping-friendly function, and requires formula and data parameters as inputs. The dot (.) indicates *whatever comes out of the pipe*, hence the selection of two variables in the line above and the specification of the data using the dots.

The RUC is also hierarchical in that it can be recast into two high-level classes of Urban and Rural:

```
social %>%
  select(oac, ruc11_code) %>%
  # recode the classes
  mutate(UR = ifelse(str_detect(ruc11_code , "A1|B1|C1|C2"),
                     "Urban","Rural")) %>%
  # select the UR and OAC variables and pipe to xtabs
  select(oac, UR) %>% xtabs( ~ ., data = .)
```

```
##                                               UR
## oac                                     Rural Urban
##     Cosmopolitan student neighbourhoods      4  1134
##     Countryside living                    3807   574
##     Ethnically diverse professionals       164  5042
##     Hard-pressed communities                95  4733
##     Industrious communities               1396  5773
##     Inner city cosmopolitan                  0  2170
##     Multicultural living                     0  3838
##     Suburban living                        738  5285
```

It could also be recast into **Sparse** and **Non-sparse** classes:

```
social %>%
  select(oac, ruc11_code) %>%
  # create the summaries
  mutate(SP = ifelse(str_detect(ruc11_code , "1"),
                  "Non-Sparse","Sparse")) %>%
  # select the SP and OAC variables and pipe to xtabs
  select(oac, SP) %>% xtabs( ~ ., data = .)
##                                               SP
## oac                                Non-Sparse Sparse
##     Cosmopolitan student neighbourhoods    1131      7
##     Countryside living                     3960    421
##     Ethnically diverse professionals       5202      4
##     Hard-pressed communities               4802     26
##     Industrious communities                7012    157
##     Inner city cosmopolitan                2170      0
##     Multicultural living                   3838      0
##     Suburban living                        6019      4
```

The results of the two code snippets above could be expanded by commenting out the `select` operation to show the detail beneath the high-level groupings and *how* each rural and urban class interacts with each OAC class within the groupings.

A classification of unemployment can be applied to the LSOA unemployed rates and linked to OAC class. The code below splits the data over the 25th, 50th and 75th percentiles to create four classes of unemployment: 1 is within the 25th (lowest) percentile, 4 is in the highest percentile. Then it evaluates how the OAC classes relate to the distribution of the unemployed variable:

```
social %>%
  mutate(`Employment quantile` = ntile(unemployed, 4)) %>%
```

```
select(oac, `Employment quantile`) %>%
xtabs( ~ ., data = .)
```

```
##                                            Employment quantile
## oac                                         1    2    3    4
##    Cosmopolitan student neighbourhoods     234  322  357  225
##    Countryside living                     2542 1412  410   17
##    Ethnically diverse professionals       1318 2028 1633  227
##    Hard-pressed communities                  6   71  977 3774
##    Industrious communities                 750 2256 3153 1010
##    Inner city cosmopolitan                 218  384  733  835
##    Multicultural living                      8  158 1076 2596
##    Suburban living                        3613 2057  349    4
```

Tables can also be visualised using a heatmap (as was done in Figure 5.8) to show the relative frequency of interaction between the two classes. The code for generating the graduated correspondence table is below:

```
social %>%
  select(oac, ruc11_code) %>% xtabs( ~ ., data = .) %>%
  data.frame() %>%
  ggplot(aes(ruc11_code, oac, fill= Freq)) +
  geom_tile(col = "lightgray") +
  xlab("") + ylab("") + coord_fixed() +
  scale_fill_gradient(low = "white", high = "#CB181D") +
  theme_bw()
```

Key Points

- Categorical data describe the *class* of an observation.

- Categories and classes can be examined by comparing their frequencies in tabular format and visually using column or bar charts.

- Correspondence tables provide a convenient way to summarise how two categorical variables interact.

- These can be cast into heatmaps in ggplot using the geom_tile function.

- Categories can be used to group or facet plots in order to explore how different group *numeric* properties and attributes vary (as was done for scatter plots in Figure 5.2).

- More complex summaries can be visualised using faceted radar plots.

A word on colours

This chapter has used different colours in the visualisations. Figure 5.1 generated a scatter plot using colours in hexadecimal format ("#FB6A4A" and "#DE2D26") and the code for the heatmaps defined a colour ramp from a named colour ("white") to a hexadecimal one ("#CB181D"). This starts at the different ways that colours can be specified. There are three basic approaches.

First, named colours can be used. These are listed as follows using both British and American spellings:

```
colours()
colors()
```

Second, they can be specified in RGB (red–green–blue) format. The code below first creates RGB values for some named colours:

```
rgb_cols = col2rgb(c("red", "darkred", "indianred", "orangered"))
rgb_cols

##        [,1] [,2] [,3] [,4]
## red     255  139  205  255
## green     0    0   92   69
## blue      0    0   92    0
```

Third, they can be in hexadecimal format:

```
rgb(rgb_cols, maxColorValue = 255)
## [1] "#FF8BCD" "#00005C" "#00005C"
```

You will have noticed that different colour palettes typically use hexadecimal format such as the calls to brewer.pal above:

```
brewer.pal(11, "Spectral")
brewer.pal(9, "Reds")
```

The named and hexadecimal formats can be used directly with ggplot, while the RGB format needs to be converted as demonstrated in the code below:

```
col_name = "orangered1"
col_rgb = t(col2rgb(col_name)/255)
col_hex = rgb(col_rgb, maxColorValue = 255)
# named colour
ggplot(data = social, aes(x = unemployed, y = noqual)) +
    geom_point(alpha = 0.1, size = 0.7,
               colour = col_name, shape = 1) +
```

(Continued)

```
  theme_minimal()
# hexadecimal colour
ggplot(data = social, aes(x = unemployed, y = noqual)) +
   geom_point(alpha = 0.1, size = 0.7,
              colour = rgb(col_rgb), shape = 1) +
   theme_minimal()
# RGB colour with conversion
ggplot(data = social, aes(x = unemployed, y = noqual)) +
   geom_point(alpha = 0.1, size = 0.7,
              colour = rgb(col_rgb), shape = 1) +
   theme_minimal()
```

The code below generates a heatmap of 64 of the 657 named colours:

```
set.seed(91234)
index = sample(1:length(colours()), 64)
# select colours
cols <- colours()[index]
# create a matrix of RGB values for each named colour
rgbmat <- t(col2rgb(cols))
# turn into hex formats
hexes <- apply(rgbmat, 1, function(x)
   sprintf("#%02X%02X%02X",x[1],x[2],x[3]))
# create backgrounds - foreground either black or white depending
# on lightness of colour
textcol <- ifelse(rowMeans(rgbmat) > 156, 'black', 'white')
# now create a data frame of cells with foreground/background colour
# spec text size
text_size = round(30/str_length(cols), 0)
df = data.frame(expand_grid(x = 1:8, y = 1:8), colours = cols,
                background = hexes, colour = textcol,
                text_size = text_size)
# apply to a plot
ggplot(df, aes(x, y, label = as.character(colours),
               fill = as.character(background),
               col = as.character(colour))) +
   geom_tile(col = "lightgray") +
   #geom_label(label.size = 0.05, col = as.character(df$colour))+
   geom_text(size = df$text_size, col = as.character(df$colour))+
   scale_fill_identity() +
   theme_minimal() + xlab("") +ylab("")+
   theme(axis.line=element_blank(),
         axis.text=element_blank(),
         axis.ticks=element_blank())
```

5.7 TEMPORAL TRENDS: SUMMARISING DATA OVER TIME

Much data analysis is concerned with temporal trends: how responses vary at different times of the day, on different days, at different periods in the year and across different years. This section outlines a visual trend analysis.

A fundamental element of any temporal data analysis or visualisation is to ensure that the data are in a date format. To do this we use the in-memory prescriptions as this includes the month of prescribing in 2018.

Let us examine a hypothetical case of seasonal affective disorder which can result in depression associated with reduced daylight and is sometimes treated with selective serotonin reuptake inhibitors (SSRIs). These are antidepressants and are believed to be most effective when taken at the start of winter before symptoms appear, and continued until spring. SSRIs correspond to the BNF code starting with 04030. The data can be filtered for this BNF code:

```
# extract data
prescriptions %>% filter(str_starts(bnf_code, "04030")) %>% head()
```

The next step is to think about dates. The as.Date and related functions are key for manipulating dates in R. The code below attaches '2018' for the year and '01' as a nominal day to make a formal date attribute to each of the objects in the filtered data, building on the code above:

```
# add date
prescriptions %>% filter(str_starts(bnf_code, "04030")) %>%
    mutate(date = as.Date(paste0("2018-", month, "-01"))) %>% head()
```

Finally, the results can be passed to ggplot:

```
prescriptions %>%
    # extract data
    filter(str_starts(bnf_code, "04030")) %>%
    # add date
    mutate(Date = as.Date(paste0("2018-", month, "-01"))) %>%
    # group by data and summarise
    group_by(Date) %>% summarise(Count = n()) %>%
    # and plot the line with a trend line
    ggplot(aes(x=Date, y=Count)) + geom_line() +
    geom_smooth(method = "loess")
```

Note the use of the group_by and summarise within the piped commands above.

This can be extended to compare different trends for different groups of drugs. The code below compares respiratory prescriptions associated with asthma (BNF code 03010). Notice the use of the nested `ifelse` within the call to `mutate` followed by the use of `filter` to extract the two sets of prescriptions and the grouping by two variables as in Figure 5.13:

```
prescriptions %>%
    # extract data
    mutate(Condition =
        ifelse(str_starts(bnf_code , "04030"), "SAD",
        ifelse(str_starts(bnf_code , "03010"),"Asthma","Others"))) %>%
    filter(Condition != "Others") %>%
    # add date
    mutate(Date = as.Date(paste0("2018-", month, "-01"))) %>%
    # group by date and summarise
    group_by(Date, Condition) %>% summarise(Count = n()) %>%
    ungroup() %>%
    # and plot the line with a trend line
    ggplot(aes(x=Date, y=Count, colour = Condition)) + geom_line() +
        geom_smooth(method = "loess") + theme_bw() +
        scale_color_manual(values = c("#00AFBB", "#E7B800"))
```

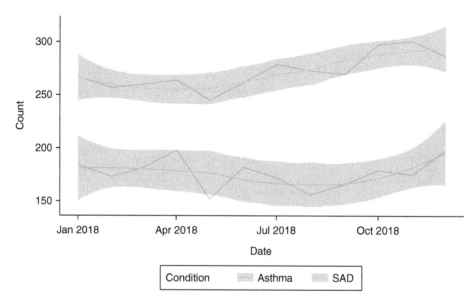

Figure 5.13 Time series plots for prescriptions associated with seasonal affective disorder (SAD) and asthma

Database version

The database version of the above is slightly different due to nuances in the way that dplyr interfaces with SQL using the DBI package. The code below connects to the full prescriptions data table in prescribing.sqlite that was introduced in Chapter 4. You should connect to this database. Note that you may have to change your working directory to the folder location of the 11.2 GB file you downloaded in Chapter 4.

```
library(RSQLite)
db <- dbConnect(SQLite(), dbname="prescribing.sqlite")
```

One of the issues with the dplyr database interface is that it does not translate the paste0 function as above, so this is moved to the ggplot part of the code, and we have to use a different pattern matching than str_detect (see Chapter 4). Also note that in the extended plot, the selection and labelling have a different syntax. The code below takes some time to run because it summarises data from 120 million records:

```
tbl(db, "prescriptions") %>%
  # extract data
  filter(bnf_code %like% '04030%') %>%
  # group by date and summarise
  group_by(month) %>%
  summarise(Count = n()) %>%
  ungroup() %>%
  # plot
  ggplot(aes(x=as.Date(paste0("2018-", month, "-15")), y=Count)) +
    geom_line() + xlab("Date") + geom_smooth(method = "loess")
```

This can be extended as before, although the results are a bit underwhelming - the monthly trends are not visible when the counts are on the same y-scale. Try uncommenting the line with facet_grid near the bottom:

```
tbl(db, "prescriptions") %>%
  # extract data
  filter(bnf_code %like% '04030%' | bnf_code %like% "03010%") %>%
  mutate(Condition = if_else(bnf_code %like% '04030%', "SAD",
"Asthma")) %>%
  # group by date and condition and summarise
  group_by(month, Condition) %>%
  summarise(Count = n()) %>%
```

(Continued)

```
ungroup() %>%
# plot
ggplot(aes(x=as.Date(paste0("2018-", month, "-15")),
           y=Count, colour = Condition)) +
  geom_line() + geom_smooth(method = "loess") +
  theme_bw() + xlab("Date") +
  facet_wrap(~Condition, nrow = 2, scales = "free") +
  scale_color_manual(values = c("#00AFBB", "#E7B800"))
```

The database can be closed:

```
dbDisconnect(db)
```

Key Points

- Trend analysis in ggplot requires temporal data to be in date format, created using the as.Date function.

5.8 SPATIAL EDA

The final section of this examination of EDA with ggplot describes how it can be used with spatial data. The tmap package (Tennekes, 2018) was formally introduced in Chapter 2, with further examples in Chapter 3, and it was used to generate maps of the database query outputs in Chapter 4. However, the ggplot2 package can also be used to visualise spatial data. This section provides some outline examples of how to do this with point and area data.

The code below subsets the lsoa_sf layer for Nottingham and links the result to the LSOA prescribing attributes in lsoa_result:

```
# define a clip polygon
ymax = 53.00; ymin = 52.907; xmin = -1.240; xmax = -1.080
pol = st_polygon(list(matrix(c(xmin,ymin,xmax,ymin,
                               xmax,ymax,xmin,ymax,xmin,ymin),
                             ncol=2, byrow=TRUE)))
# convert polygon to sf object with correct projection
pol = st_sfc(st_cast(pol, "POLYGON"), crs = 4326)
# re-project it
pol = st_transform(pol, 27700)
# clip out Nottingham from lsoa_sf making the geometry valid
nottingham = lsoa_sf[pol,] %>% inner_join(lsoa_result) %>%
  st_make_valid()
```

It is also possible to visualise or map spatial data using the ggplot2 package. The code below plots the nottingham dataset introduced above, but, rather than using tmap (or qtm), uses the ggplot2 package, with the results shown in Figure 5.14:

```
ggplot(nottingham) + geom_sf()
```

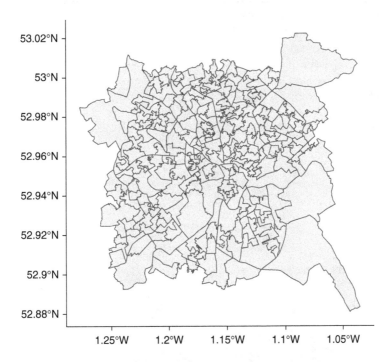

Figure 5.14   The map of the LSOAs in Nottingham created using ggplot

The geom_sf function adds a geometry stored in an sf object to the plot. It looks for an attribute called geom or geometry in the sf object. On occasion you may need to specify this as in the code snippet below to reproduce Figure 5.14:

```
ggplot(nottingham) + geom_sf(aes(geometry = geom))
```

As with tmap and other ggplot operations, colour can be specified directly or for a variable, using fill or colour depending on the what is being shaded in the plot:

```
# directly
ggplot(nottingham) +
  geom_sf(aes(geometry = geom), colour = "darkgrey", fill = "azure")
```

```
# using a continuous variable and specify a thin line size
ggplot(nottingham) +
  geom_sf(aes(geometry = geom, fill = cost_pp), size = 0.1) +
  scale_fill_viridis_c(name = "Opioid costs \nper person")
```

It is also possible to add scale bars and north arrows using some of the functions in the ggspatial package, and to control the appearance of the map using both predefined themes (such as theme_bw) and specific theme elements. Examples of some of these are shown in the code below to create Figure 5.15:

```
ggplot(nottingham) +
  geom_sf(aes(geometry = geom, fill = cost_pp), size = 0.1) +
  scale_fill_viridis_c(name = "Opioid costs \nper person") +
  annotation_scale(location = "tl") +
  annotation_north_arrow(location = "tl", which_north = "true",
          pad_x = unit(0.2, "in"), pad_y = unit(0.25, "in"),
          style = north_arrow_fancy_orienteering) +
  theme_bw() +
  theme(panel.grid.major = element_line(color = gray(.5),
                                        linetype="dashed",size=0.5),
        panel.background = element_rect(fill="white"))
```

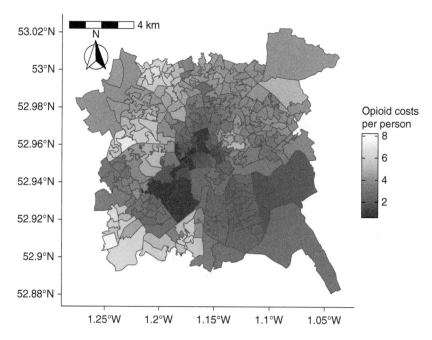

Figure 5.15 Map of limiting long-term illness in Nottingham using ggplot with cartographic enhancements

Finally, as with tmap, multiple data layers can be included in the map. The code below creates and then adds the locations and names of some randomly selected LSOAs to the map:

```
# select locations
set.seed(13)
labels.data = nottingham[sample(1:nrow(nottingham), 10), ]
labels.data = cbind(labels.data,
                    st_coordinates(st_centroid(labels.data)))
# create ggplot
ggplot(nottingham) +
  geom_sf(aes(geometry = geom), fill = "white", size = 0.1) +
  geom_point(data = labels.data, aes(x=X, y=Y),
             colour = "red") +
  geom_text(data = labels.data, aes(x=X, y=Y, label=lsoa_id),
    fontface = "bold", check_overlap = T) +
  theme_bw() +
  theme(axis.title = element_blank())
```

It is possible to use ggplot and tmap to visualise analyses of *spatial* structure in spatial data variables.

The key thing that we are interested in, which cannot be directly determined using the *aspatial* bivariate and multivariate approaches (correlations, correspondence tables, etc.), is the locations *where*, for example, correlations were high or low. Some approaches are outlined for determining similar properties but with a spatial flavour in the remainder of this section.

The code below does a similar thing to what was done in some of the ggpairs plots above that partitioned the social data around median unemployment. This time it is done using the nottingham spatial object with the opioid prescribing results as presented in Figure 5.16. This shows that there are clear spatial trends, with a classic doughnut effect around the city centre. Both tmap and ggplot approaches are shown:

```
# tmap
p1 = nottingham %>%
  mutate(cost_high = ifelse(cost_pp > median(cost_pp),
                            "High","Low")) %>%
  tm_shape() +
    tm_graticules(ticks = FALSE, col = "grey") +
    tm_fill("cost_high", title = "Cost pp.",
            legend.is.portrait = FALSE) +
    tm_layout(legend.outside = F,
              legend.position = c("left", "bottom"))
# ggplot
p2 = nottingham %>%
```

```
mutate(cost_high = ifelse(cost_pp > median(cost_pp),
                       "High","Low")) %>%
ggplot() +
  geom_sf(aes(geometry = geom, fill = cost_high), size = 0.0) +
  scale_fill_discrete(name = "Cost pp.") +
  theme_bw() +
  theme(panel.grid.major = element_line(
      color=gray(.5), linetype="dashed", size=0.3),
        panel.background = element_rect(fill="white"),
        legend.position="bottom",
        axis.text = element_text(size = 6))
# plot together using grid
library(grid)
# clear the plot window
dev.off()
pushViewport(viewport(layout=grid.layout(2,1)))
print(p1, vp=viewport(layout.pos.row = 1))
print(p2, vp=viewport(layout.pos.row = 2))
```

Rather than a binary approach to folding the data (e.g. around a median as above), another way is to use that as an inflection point and use colour to indicate the degree to which individual LSOAs are above or below it. Divergent palettes can help as in Figure 5.17. Notice the placement of the legend:

```
ggplot(nottingham) +
  geom_sf(aes(geometry = geom, fill = cost_pp), size = 0.0) +
  scale_fill_gradient2(low="#CB181D", mid = "white", high="#2171B5",
                    midpoint = median(nottingham$cost_pp),
                    name = "Unemployment\n around median") +
theme_bw() +
theme(axis.line=element_blank(),
      axis.text=element_blank(),
      axis.ticks=element_blank(),
      legend.position = "bottom",
      legend.direction = "horizontal")
```

The choice between ggplot and tmap approaches may relate to the user's norms about maps, which many people believe should *always* be accompanied by a scale bar and a north arrow. Classic map production with common cartographic conventions such as legends, scale bars and north arrows is readily supported by tmap, and although possible with ggplot2 is less directly implemented in the latter. A comprehensive treatment of the use of the tmap and ggplot2 packages to map spatial data is given in Brunsdon and Comber (2018).

Chapter 8 describes a number of alternative visualisations of spatial data using hexbins and different kinds of cartograms.

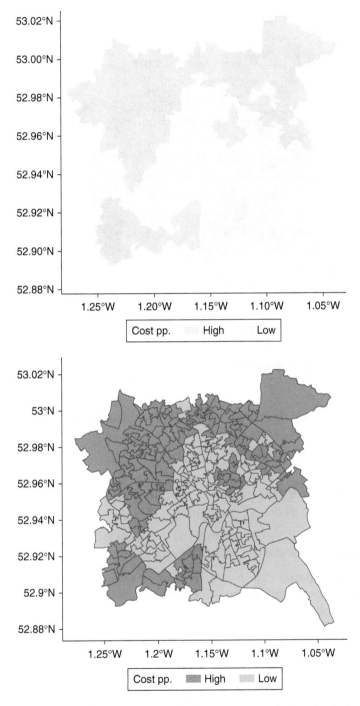

Figure 5.16 LSOAs with high and low areas of opioid prescribing using (top) tmap and (bottom) ggplot approaches

Unemployment around median

2   4   6   8

Figure 5.17   LSOA opioid prescribing costs, shaded above (blue) and below (red) the median rate

Key Points

- Spatial EDA (mapping!) can be undertaken with ggplot as well as tmap using the geom_sf function.

- This is supported by the ggspatial package which contains functions for map embellishments (scale bars, north arrows, etc.).

- Continuous spatial data properties can be visualised by partitions (e.g. around the mean or median) and supported by divergent palettes.

5.9 SUMMARY

R/RStudio provides a fertile environment for the development of tools for visualising data, data attributes and their statistical and distributional relationships.

It is impossible to capture the full dynamics of this information environment: new webpages, blogs, tutorials and so on appear constantly. In this context the aims of this chapter are not to provide a comprehensive treatment of ggplot2 or of all the different mapping options for spatial data. However, we have illustrated many generic ggplot visualisation techniques with some implicit design principles, highlighting why the ggplot2 package has become one of the default packages for data visualisation.

A range of different types of visualisation have been used to illustrate specific points in previous chapters, including lollipop charts, scatter plots, boxplots, histograms, density plots and pairwise correlation plots. This chapter has sought to bring these together, identifying some common themes, for example the use of layered approaches to graphics and maps, and highlighting particular techniques. We have not by definition covered everything – bar charts, parallel plots, calendar heatmaps, for example. The potential for visualisation refinement with ggplot2 and its layered approach is endless. The code snippets in this chapter illustrate the syntax of ggplot and provide the reader with baseline skills to allow them to understand how other plot types could be implemented.

A further point is that ggplot2 provides the engine for much of the development of visualisation tools in R. It has been adapted and extended in many other packages – have a look at the packages beginning with gg on the CRAN website (https://cran.r-project.org/web/packages/available_packages_by_name.html), some of which have been used in this and other chapters:

- GGally was used to produce the ggpairs plots but has other very useful functions such as ggcoef() which plots regression model coefficients. Consider the aim to model opioid costs (cost_pp) from the other socio-economic variables in the social dataset using regression. The code below creates a model of cost_pp, prints out the model coefficients and plots them with 95% confidence intervals using the ggcoef() function:

```
df = nottingham %>% st_drop_geometry() %>% inner_join(social) %>%
    select(-c(lsoa_id, population, employed, tot_cost,
             ruc11_code, ruc11, oac, oac_code))
regression.model <- lm(cost_pp ~ ., data = df)
round(summary(regression.model)$coefficients, 4)
ggcoef(regression.model, errorbar_height=.2,
       color ="red", sort="ascending") + theme_bw()
```

- ggspatial provides a small set of useful tools for cartographic enhancements for ggplot visualisations of spatial data, including scale and error bars. Again these are precompiled snippets of ggplot codes to produce map embellishments, which are evident if the code is examined:

```
annotation_north_arrow
```

Because of the flexibility and layer structure of graphics and visualisations with ggplot2, the potential development of new visualisation tools and routines is endless. Readers interested in developing these aspects are encouraged to explore them further through the ggplot book (Wickham, 2016) which can also be compiled for free, as well ggplot cheat sheets (see https://ggplot2.tidyverse.org/). There are also large amounts of resources online via repositories, blogs and online tutorials which can easily be found through internet searches (e.g. http://r-statistics. co/Top50-Ggplot2-Visualizations-MasterList-R-Code.html).

More generally, there are some excellent texts on data visualisation generally and specifically using R. Many of them have some kind of online or bookdown version that is free to use. Good general texts with a focus on the principles of visualisation include Kirk (2016) and Tufte and Robins (1997), while Tufte (2001) provides a touchstone (i.e. everyone should read it).

Various R-specific texts provide insight into specific packages as well as more general guidance about how to use different visualisation techniques: Wickham (2016), Chang (2018) and Healy (2018); see also Kabacoff (2018).

REFERENCES

Brunsdon, C. and Comber, L. (2018) *An Introduction to R for Spatial Analysis and Mapping* (2nd edn). London: Sage.

Chang, W. (2018) *R Graphics Cookbook: Practical Recipes for Visualizing Data.* Sebastopol, CA: O'Reilly Media.

Cleveland, W. S. (1979) Robust locally weighted regression and smoothing scatterplots. *Journal of the American Statistical Association*, 74(368), 829–836.

Healy, K. (2018) *Data Visualization: A Practical Introduction.* Princeton, NJ: Princeton University Press.

Kabacoff, R. (2018) Data Visualization with R. https://rkabacoff.github.io/datavis/.

Kirk, A. (2016) *Data Visualisation: A Handbook for Data Driven Design.* Thousand Oaks, CA: Sage.

Tennekes, M. (2018) tmap: Thematic maps in R. *Journal of Statistical Software*, 84(6), 1–39.

Tufte, E. R. (2001) *The Visual Display of Quantitative Information*, Vol. 2. Cheshire, CT: Graphics Press.

Tufte, E. R. and Robins, D. (1997) *Visual Explanations*. Cheshire, CT: Graphics Press.

Wickham, H. (2010) A layered grammar of graphics. *Journal of Computational and Graphical Statistics*, 19(1), 3–28.

Wickham, H. (2016) *ggplot2: Elegant Graphics for Data Analysis.* Cham: Springer.

Wilkinson, L. (2012) The grammar of graphics. In J. E. Gentle, W. Härdle and Y. Mori (eds), *Handbook of Computational Statistics* (pp. 375–414). Berlin: Springer.

6

MODELLING AND EXPLORATION OF DATA

6.1 OVERVIEW

Why do we model at all? This may sound like a strange question, but the answers are surprisingly diverse. We would argue that there is no simple answer, and that the reasons depend greatly on context.

This chapter develops a series of models to illustrate some of the core modelling paradigms, with a key focus on the development of models and testing their assumptions. It illustrates two key types of inference – classical and Bayesian – and suggests how these can be used to calibrate models, test hypotheses or provide predictions. The main aim of this chapter is to provide some guidelines for thinking about models, and how they relate to data.

You will need to load the following packages in this chapter:

```
library(tidyverse)
library(dbplyr)
library(biglm)
library(DBI)
library(RSQLite)
library(biglm)
library(lubridate)
library(sf)
library(tmap)
library(spdep)
```

You will also need to download some house price and socio-economic data for Liverpool. Details of this are in Section 6.3.

In the physical sciences, there is quite often a very rigorous derivation of the relationship between a set of quantities, expressed as an equation. However, measurements may be necessary to calibrate some physical constants in the equation. An example of this is van der Waals' equation of state for fluids (van der Waals, 1873):

$$\left(P + a\left(\frac{n}{V}\right)^2\right)\left(\frac{V}{n} - b\right) = RT \tag{6.1}$$

This is a modification of the *ideal gas law*:

$$PV = nRT \tag{6.2}$$

where P is pressure, V is the volume occupied by the container of the fluid, R is the universal gas constant,[1] T is temperature based on an absolute zero scale (e.g. kelvin), and n is the amount in moles (molecular weight in grams) of the fluid. The two remaining quantities, a and b, are constants for a given fluid – and are universally zero in the ideal gas law. Since it is possible to measure or calculate V, P, n and T with good accuracy, and R is known, it is possible to take observations over a variety of values for a given fluid, and then use non-linear regression techniques to obtain estimates for a and b.

The point here is not to provide a tutorial on physical chemistry, but to spotlight a certain approach to modelling. In the basic (so-called 'ideal') gas law, molecules are presumed to be zero-dimensional points (occupying no volume) and intermolecular attraction is also ignored. Van der Waals considers the implication of dropping these two assumptions, and in each case linking these with an explicit and rigorously thought-through mathematical modification. Essentially the coefficient a allows for intermolecular attraction, and the coefficient b allows for the volume occupied by each molecule – these will vary for different fluids. It is unlikely that anyone would pick that equation out of thin air – the model is the outcome of quite precise mathematical (and physical) reasoning.

However, this is not the only way that models are produced. Quite often the objective is to understand processes for which data can be collected, and it is expected that certain variables will be connected, but there is no clear pathway (as with the van der Waals example) to deriving a mathematical model such as equation (6.1). Thus, it may be expected that house prices in an area are related to, say, site properties (i.e. properties of the house itself), neighbourhood properties and environmental properties. These are essentially the assumptions made by Freeman (1979). Suppose we had a number of measurements relating to each of these, respectively named S_1, ..., S_{k1}, N_1, ..., N_{k2} and E_1, ..., E_{k3}. If P is a house price then we might expect P to depend on the other characteristics. Generally we could say

$$P = f(S_1, ..., S_{k1}, N_1, ..., N_{k2}, E_1, ..., E_{k3}) \tag{6.3}$$

Here, it is stated that some kind of function links the quantities, but it is just referred to as f – a general function. A further step towards reality is to state

---

[1] Approximately 8.314 J mol^{-1} K^{-1} – the product of Boltzmann's constant and Avogadro's number.

$$P = f(S_1, \ldots, S_{k1}, N_1 \ldots N_{k2}, E_1 \ldots E_{k3}, \epsilon) \tag{6.4}$$

where ϵ is a stochastic (random) element. This is generally recognised now as a *hedonic price equation*. Here, a great deal of thought has still gone into the derivation of this model, building on Rosen (1974), Freeman (1974) and Lancaster (1966); however, there is no clear guidance as to the form that f should take. For example, Halvorsen and Pollakowski (1981) argue that '[the] appropriate functional form for a hedonic price equation cannot in general be specified on theoretical grounds' and that it should be determined on a case-by-case basis. They go on to produce a quite flexible model (involving Box–Cox power transformations of the predictors of house price choice, and linear and quadratic terms of these), where statistical tests may be used to decide whether some simplifications can be made. However, this model is justified essentially by its flexibility, rather than direct derivation in the style of van der Waals. In a hundred years a progression can be seen in which the set of applications for quantitative modelling has widened (from the physical to the economic), but the certainty in the form of the model is reduced.

This trend has perhaps accelerated in more recent decades, to a situation where machine learning is proposed as a means of finding a functional form (without necessarily stating the function itself) and very little consideration of the underlying process generating the data is considered. Indeed, this is considered by some (e.g. Anderson, 2008) to be a major advantage of the approach.

However, while the situation has accelerated, opinion has diversified, and others would argue strongly against a theory-free approach – see Kitchin (2014), Brunsdon (2015) or Comber and Wulder (2019) for examples. In fact there is a spectrum of opinion on this, perhaps leading to a diverse set of approaches being adopted. As well as work on 'black box' models of machine learning and model fitting, a number of new approaches have emerged from the statistical community – for example, generalised additive models (Hastie and Tibshirani, 1987) where a multivariate regression of the form

$$y = f_1(x_1) + f_2(x_2) + \ldots + f_m(x_m) + \epsilon \tag{6.5}$$

is proposed, in which $f_1, \ldots, f_k$ are arbitrary functions and ϵ is a Gaussian error term with zero mean. The framework can be adapted to other distributions in the exponential family, such as Poisson or binomial, with appropriate transformations of y. Although this form is not entirely general, it does have the property discussed earlier that it is quite flexible – providing flexibility in the situation where there is no obviously functional form for the model. There are statistical tests proposed to evaluate whether certain terms should be included in the model, or whether some of the arbitrary fs could be replaced by a linear term. Another reason why attention is paid to models with a range of statistical sophistication is the advent of 'big data'. More flexible models tend to require a greater degree of computational intensity, and sometimes they do not scale well when the size of datasets becomes large. It may be more practical to use a simple linear regression model – if it is a reasonable approximation of reality in the situation where it is applied – rather than a more complicated model.

Also, more flexible models have more parameters to estimate, and for equivalent-sized data samples they will be more prone to sampling variability. This may be seen as a price worth paying if there is no reason to adopt a particular model, but if there is a strong theoretical argument that a specific mathematical form should apply then there is no need to use a more general – and more variable – approach. Sometimes a balance has to be struck between an inaccurate model and an unreliably calibrated model.

Finally, the term 'inaccurate' is deliberately used here, rather than 'wrong'. There are degrees of inaccuracy. Newton's laws predict planetary motion to an extremely high degree of accuracy, despite being proven 'wrong' by Einsteinian relativity. It took a very long time and a major theoretical re-evaluation to identify a need for revision.

Key Points

- Some models are directly derived from considered and precise mathematical and physical reasoning about the process being considered (as in the van der Waals example).

- Others are constructed through an understanding of the process being considered, and this may be subjective (as in the house price example).

- The applications for quantitative modelling have widened from the physical sciences to the economic and others, but the certainty in the form of the model is less.

- This has been exacerbated by the increase in machine learning applications which are good at finding an appropriate functional form (they are flexible), but with very little consideration of the underlying process.

6.2 QUESTIONS, QUESTIONS

Much of the purpose of modelling is to answer questions – for example, 'By how much would the average house price in an area increase if it was within 1 km of a woodland area?', 'What is the pressure of 1 mole of hydrogen gas contained in a volume of 500 cm^3 at a temperature of 283 K?', or 'Is there a link between exposure to high-volume traffic and respiratory disease?'. Questions may be answered with varying degrees of accuracy, and varying degrees of certainty. The gas pressure question can be answered accurately and with a very high degree of confidence

(49.702 atm). The others less so – for the house price case even the question asked only for the *average* house price, given a number of predictors. In reality it might also be useful to specify some kind of range – or at least an indicator of variability. The last question implies a yes/no response. One could make such a response, but realistically this would have some probability of being wrong. It would be helpful if that probability could be estimated.

Perhaps something to be deduced from these comments is that there is a need – in many cases – not only to answer a question, but also to obtain some idea of the reliability of the answer. In some cases this might be relatively informal, while in others a more formal quantified response can be considered – for example, a number range, a probability distribution for an unknown quantity, or a probabilistic statement on the truth of an answer. The latter will also require data to be collected and provided. Part of this problem can be addressed by statistical theory and probability theory, but a number of other issues need to be addressed as well – for example, the reliability of the data, or the appropriateness of the technique used to provide numerical answers. Good data science should take all of these into account. In the following subsections, a number of these issues will be addressed, with examples in R. These examples do not present the *only* way to answer questions, but are intended to give examples of thinking that *could* be of help. If readers can come up with effective critiques of the approaches, one intended aim of this book will be achieved!

6.2.1 Is this a fake coin?

Probability-based questions are possibly some of the simplest to answer, particularly with the use of R. Here, there is a relatively small amount of data, but similar approaches using larger datasets are possible. The key idea here is to think about how the question can be answered in probability terms.

Suppose you are offered a choice of two coins, one known to be genuine, and one a known fake. The coin you take is flipped five times, and four heads are obtained. We also know that for this kind of coin, genuine versions are 'fair' (in that the probability of obtaining a head is 0.5). For the fake coin, however, the probability is 0.6. Based on the result of coin flipping above, is the coin a fake?

One problem here is that of *equifinality* (Von Bertalanffy, 1968) – coins with head probability of either 0.5 or 0.6 *could* give four heads out of five. Different processes (i.e. flipping a genuine coin versus flipping a fake coin) could lead to the same outcome. However, it may be that the outcome is less likely in one case than another – and in turn this could suggest how the likelihoods for the two possible coin types differ. Here it is equally likely that we have a fair or fake coin – due to the initial choice of coins offered. The probability that the coin is fair, given the outcome, could be calculated analytically. However, since part of the idea here is to consider the possibility of more complicated outcomes, an experimental approach will be used, as set out here:

1. Generate a large number (n) of simulated sets of five coin flips. To represent the belief that the coin is equally likely to be fake or genuine, select the chance of getting heads to be either 0.5 or 0.6 with equal probability.

2. Store the outcomes (number of heads), and whether the probability was 0.5 or 0.6.

3. Select out those sets of coin flips with four heads.

4. Compute the experimental probability of the coin being genuine (given the result) by finding the proportion of 0.5-based results.

This is done below. The seed for the random number generator is fixed here, for reproducibility. probs contains the probability for each set of coin flips. The results are obtained using the rbinom function, which simulates random, binomially distributed results – effectively numbers of heads when the coin is flipped five times. n_tests here is 1,000,000, allowing reasonable confidence in the results. The filtering and proportion of genuine/fake coin outcomes are computed via dplyr functions:

```
library(tidyverse)
set.seed(299792458)
n_tests <- 1000000
probs <- sample(c(0.5,0.6),n_tests,replace=TRUE)
results <- tibble(Pr_head=probs,Heads=rbinom(n_tests,5,probs))
results %>% filter(Heads==4) %>%
   count(Pr_head) %>%
   mutate(prop=n/sum(n)) %>%
   select(-n)
## # A tibble: 2 x 2
##    Pr_head prop
##      <dbl> <dbl>
## 1      0.5 0.377
## 2      0.6 0.623
```

So, on the basis of this experiment it looks as though the odds of the chosen coin being a fake (i.e. a coin with a heads probability of 0.6) are about 2 : 1.

Now, suppose instead of being offered a choice of two coins, this was just a coin taken from your pocket. There are fake coins as described above in circulation (say, 1%) so the chance of the probability of getting a head being 0.6 is now just 0.01, not 0.5. The code can be rerun (below), adapting the sample function with the probs parameter:

```
set.seed(299792458)
n_tests <- 1000000
```

```
probs <- sample(c(0.5,0.6),n_tests,replace=TRUE,prob=c(0.99,0.01))
results2 <- tibble(Pr_head=probs,Heads=rbinom(n_tests,5,probs))
results2 %>% filter(Heads==4) %>%
  count(Pr_head) %>%
  mutate(prop=n/sum(n)) %>%
  select(-n)
## # A tibble: 2 x 2
##     Pr_head    prop
##       <dbl>   <dbl>
## 1       0.5 0.983
## 2       0.6 0.0172
```

There is still some chance of the coin being fake, but now it is much less than before (about 2%).

In fact these calculations could have been dealt with analytically. The coin flips follow a binomial distribution, so that if the probability of a head is θ, then the probability of m heads out of n flips is

$$P(m \mid \theta) = \binom{m}{n} \theta^m (1-\theta)^{n-m} \tag{6.6}$$

When $m = 4$, $n = 5$, and $\theta = 0.5$, for the true coin we have $P(5 \mid 0.5) = 5/32$. This binomial is derived using factorials, denoted by $n!$ for the factorial of n:

$$\binom{n}{m} = \frac{n!}{m!(n-m)!} = \frac{120}{24} = 5 \tag{6.7}$$

In R the probability of m heads out of n flips for the true coin be calculated as follows:

```
factorial(5)/factorial(4)*(factorial(5-4)) *(0.5^4)*(1-0.5)^(5-4)
## [1] 0.15625
```

And this can be converted to a fraction using the fractions function in the MASS package. This attempts to convert a decimal result to a fraction – sometimes useful when working with probabilities:

```
MASS::fractions(factorial(5)/factorial(4)*(factorial(5-4))
*(0.5^4)*(1-0.5)^(5-4))
## [1] 5/32
```

For the fake coin, with $\theta = 0.6$, the probability of m heads out of n flips is 162/625:

```
MASS::fractions(factorial(5)/factorial(4)*(factorial(5-4))
*(0.6^4)*(1-0.6)^(5-4))
```

We can then use Bayes' theorem to work out the posterior probability that a coin is fake. The basic formulation of this is the probability of the event over the sum of the probabilities of all events:

$$P(\text{Event1} \mid \text{Event2}) = \frac{P(\text{Event2} \mid \text{Event1}) \times P(\text{Event1})}{P(\text{Event2})} \qquad (6.8)$$

where

prior probability of Event1 = P(Event1),

marginal probability of Event2 = P(Event2),

posterior probability = P(Event1 | Event2),

likelihood = P(Event2 | Event1).

Here the likelihood is 0.5, the prior probability is 162/625, and the evidence is the sum of our belief: in this case there is an equal chance of getting either a fake or real coin.

Thus without doing any experimentation (i.e. when the prior probability of a coin being fake is 50%) we can determine the posterior probability of the coin being fake:

$$P(\text{fake} \mid \text{data}) = \frac{\dfrac{1}{2} \times \dfrac{162}{625}}{\dfrac{1}{2} \times \dfrac{162}{625} + \dfrac{1}{2} \times \dfrac{5}{32}} = \frac{355}{569} = 0.6239 \qquad (6.9)$$

which agrees with the experimental result to three decimal places. The calculation can be carried out in R as follows, this time using the dbinom function to generate the probabilities of m heads out of n flips for the two coins rather than the use of the long-hand factorial approach above:

```
# prior probabilities
p_fake <- 0.5
p_gen  <- 0.5
# probability of 4 heads
p_4_fake <- dbinom(4,5,0.6)
p_4_gen  <- dbinom(4,5,0.5)
```

```
# posterior probability using Bayes
p_fake_after <- p_fake * p_4_fake / (p_fake * p_4_fake + p_gen *
p_4_gen)
p_fake_after
```

```
## [1] 0.6239018
```

```
# show fraction/ratio
MASS::fractions(p_fake_after)
```

```
## [1] 355/569
```

A similar calculation for the situation where the probability of a fake is 0.01 is as follows:

```
p_fake <- 0.01
p_gen   <- 0.99
p_4_fake <- dbinom(4,5,0.6)
p_4_gen  <- dbinom(4,5,0.5)
p_fake_after <- p_fake * p_4_fake / (p_fake * p_4_fake + p_gen *
p_4_gen)
p_fake_after
```

```
## [1] 0.01648022
```

```
MASS::fractions(p_fake_after)
## [1] 576/34951
```

This gives P(fake | data) = 0.01648 – close to the experimental value but not quite as accurate. This is in part due to the fact that the number of simulations based on the fake coins is notably less (1% of all simulations rather than 50% as before).

6.2.2 What is the probability of getting a head in a coin flip?

Here, a variation on the previous problem is introduced. In this case you have a coin, but do not know what the probability of flipping and getting a head is. An experiment tossing it 20 times gives 13 heads. The question here is 'what is the probability of getting a head in a coin flip?'. As before, equifinality is an issue – if the probability of flipping a head is θ there are lots of values of θ that could give this outcome. In fact any allowable value of θ except $\theta = 0$ or $\theta = 1$ could yield this outcome. Again, the aim here is to assess the relative likelihood of possible values of θ given this outcome.

In this case, nothing is known about the chances of this coin landing on a head – so prior to the coin flipping, any value between 0 and 1 is equally likely. Again, although a theoretical approach is possible, an experimental approach is demonstrated here initially. This is similar to the previous approach, and is set out below:

1. Generate a large number (n) of simulated sets of 20 coin flips. To represent no prior knowledge of θ, for each set of flips, generate from a uniform distribution on [0, 1].

2. Store the outcomes (number of heads), and the value of θ.

3. Select out those sets of coin flips with 13 heads.

4. The distribution of θ values from these selected results represents the probability distribution for θ on the basis of the experimental outcome.

This is done in the following code. The simulation is done as before with `rbinom`, and the distribution of θ (again referred to as Pr_head) is visualised using `geom_histogram` in ggplot2 (see Figure 6.1):

```
library(ggplot2)
set.seed(602214086)
n_tests <- 1000000
probs <- runif(n_tests)
results <- tibble(Pr_head=probs,Heads=rbinom(n_tests,20,probs))
ggplot(results %>% filter(Heads==13),aes(x=Pr_head)) +
  geom_histogram(bins=40,fill='darkred')
```

This gives an idea of the likely value of θ on the basis of the 20 flips. As before, Bayes' theorem, now with a more formal notation, can be used to obtain a theoretical distribution for θ given the outcome. If the experimental result (the 13 out of 20 heads) is referred to as D then

$$P(\theta \mid D) = \frac{P(D \mid \theta)P(\theta)}{\int_\theta P(D \mid \theta)P(\theta)d\theta} \tag{6.10}$$

Here $P(\theta)$ is the uniform distribution, and $P(D \mid \theta)$ is binomial (with $m = 13$ and $n = 20$). It can then be shown that

$$P(\theta \mid m, n) = \frac{\theta^m (1-\theta)^{n-m}}{\int_0^1 \theta^m (1-\theta)^{n-m} d\theta} \tag{6.11}$$

which is the *beta* distribution with parameters $\alpha = m + 1$ and $\beta = n - m + 1$. Although it is quite a complicated task, the experimental distribution for θ given the coin flips is compared with the theoretical one graphically (Figure 6.2) using the code below. The comments should explain the methods used:

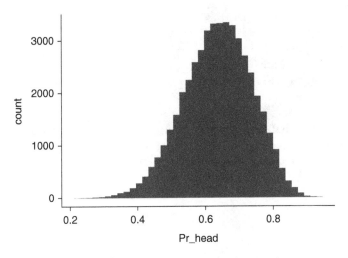

Figure 6.1 Experimental distribution of θ given coin-flipping result

```
# Here it is necessary to compute the histogram categories manually.
# 1. The following four are helper functions:
# Map each actual theta value to the bin category mid-point
cat_levels <- function(x,n) round(floor(x*n)/n + 1/(2*n),4)
# Compute the mid-points for each category
cat_centre <- function(n) round(seq(1/(2*n),1-1/(2*n),length=n),4)
# Compute the upper point for each category
cat_upper <- function(n) seq(1/n,1,length=n)
# Compute the lower point for each category
cat_lower <- function(n) seq(0,1-1/n,length=n)
# 2. Select out the results with 13 heads, map the thetas to the
# bin mid-points and count them. Label this the 'experimental'
# counts and store in 'results_h13'
results %>% filter(Heads==13) %>%
  mutate(Pcat=cat_levels(Pr_head,40)) %>%
  count(Pcat) %>%
  mutate(Type='Experimental') -> results_h13
# 3. Work out the theoretical counts and append them
n_13 <- sum(results_h13$n)
# 4. Compile a data table of the theoretical counts
tibble(n=n_13 *
             (pbeta(cat_upper(40),14,8) -
             pbeta(cat_lower(40),14,8)),
             Pcat=cat_centre(40),
             Type="Theoretical") %>%
  # 5. Join them to the experimental counts
```

```
  bind_rows(results_h13) %>%
  # Rename the columns and overwrite 'results_h13'
  rename(Count=n,Theta=Pcat) -> results_h13
# 6. Compare distributions via a plot
ggplot(results_h13,aes(x=Theta,y=Count,fill=Type)) +
  geom_col(position='dodge')
```

Here, it can be seen that the experimental approach has worked well. In particular, this is *hopefully* a good omen for those situations where an experimental approach is possible, but no theoretical analysis has been proposed.

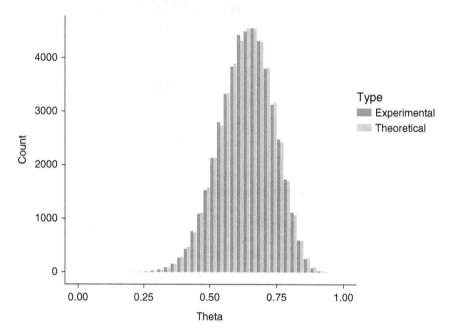

Figure 6.2   Comparison of theoretical and experimental distribution of θ given D

The question here is of the kind 'what is the value of some parameter?', rather than 'is some statement true?'. This kind of analysis could be turned into the latter kind, given that a distribution for θ has been provided. For example, suppose there was concern as to whether the coin was fair. Although with a continuous distribution it does not really make much sense to ask what the probability is that $\theta = 0.5$ *exactly*, one could decide that if it were between 0.49 and 0.51 this would be reasonable, and estimate this probability:

```
results %>%
  filter(Heads==13) %>%
  mutate(in_tol = Pr_head > 0.49 & Pr_head < 0.51) %>%
```

```
   count(in_tol) %>%
   transmute(in_tol,prob=n/sum(n))
## # A tibble: 2 x 2
##    in_tol     prob
##    <lgl>     <dbl>
## 1 FALSE     0.969
## 2 TRUE      0.0307
```

Thus, the probability that θ lies in the interval (0.49, 0.51) is just over 0.03, given the coin-flipping result.

We can also modify this analysis in the light of further information. Suppose, as before, you were offered two coins prior to flipping, a known genuine one and a known fake one, and you choose one of the two, not knowing which it was. This time you know genuine coins have $\theta = 0.5$ but do not know the value of θ for a fake. This prior knowledge of θ could be simulated by randomly generating θ as follows:

1. Simulate 'fake' or 'genuine' with respective probabilities 0.5 for each (to represent your knowledge of whether coin is fake).

2. If genuine, set θ to 0.5.

3. If fake, set θ to a uniform random number between 0 and 1.

Thus, θ has a distinct chance of being *exactly* 0.5 and an equal chance of being anywhere between 0 and 1, with uniform likelihood. The code to generate θ on this basis and simulate the coin flips is as follows:

```
set.seed(6371)
n_sims <- 1000000
theta <- ifelse(runif(n_sims) < 0.5,runif(n_sims),0.5)
results3 <- tibble(Theta=theta,Heads=rbinom(n_sims,20,theta))
```

As before, it is possible to see the distribution of θ values that lead to the observed outcome of 13 heads out of 20. Since a number of the θ values here will be *exactly* 0.5, these can be counted directly:

```
results3 %>%
   filter(Heads==13) %>%
   mutate(Hp5=Theta==0.5) %>%
   count(Hp5) %>%
   transmute(Hp5,prob=n/sum(n))


## # A tibble: 2 x 2
##    Hp5      prob
```

```
##      <lgl>   <dbl>
## 1  FALSE    0.391
## 2   TRUE    0.609
```

From this, it seems that in this case the likelihood of the coin being genuine is around 0.6. This may seem strange given the earlier results, but two things need to be considered:

- The probability of the coin actually having $\theta = 0.5$ is higher here – in the previous example no prior expectation was based on any possible value.

- In the case of a fake coin, the prior expectation of θ *exceeding* 0.5 was the same as not exceeding this value.

As before, the evidence could be viewed graphically (see Figure 6.3):

```
results3 %>%
    filter(Heads==13) %>%
    ggplot(aes(x=Theta)) + geom_histogram(bins=40,fill='navyblue')
```

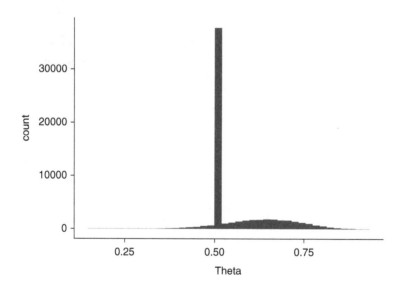

Figure 6.3 Distribution of θ given outcome of coin flipping when value is unknown for a fake

This is reasonably easy to interpret subjectively. There is a strong chance that the coin is fair, but if not then it is likely to have a θ value between around 0.4 and 0.8, peaking at about 0.65.

6.2.3 How many heads next time I flip the coin?

All of the above calculations have involved making inferences about θ for the coin being flipped, or answering logical (yes/no) questions related to this. However, data analysis and modelling are often used to make predictions. In this example, this could involve predicting the likely number of heads the *next time* the coin is flipped 20 times. One way of considering this (as before) is via simulation and repeated experimentation. By filtering out the outcomes with 13 heads (as in the above suggestion) and then running the coin-flipping experiment once for each of the experimentally selected θ values, an idea of the range of likely outcomes may be obtained. For example, on the basis of the last analysis (giving rise to `results3`) the following outcome could be predicted, and visualised as a probability distribution:

```
set.seed(271828)  # Reproducibility!
# Simulate the future coin flipping
results3 %>%
   filter(Heads==13) %>%
   mutate(`Heads
predicted`=factor(rbinom(nrow(.),20,Theta),levels=0:20)) ->
predictions
# Show predicted outcome as a histogram
predictions %>%
   ggplot(aes(x=`Heads predicted`)) +
   geom_bar(fill='firebrick') +
   scale_x_discrete(drop=FALSE)
```

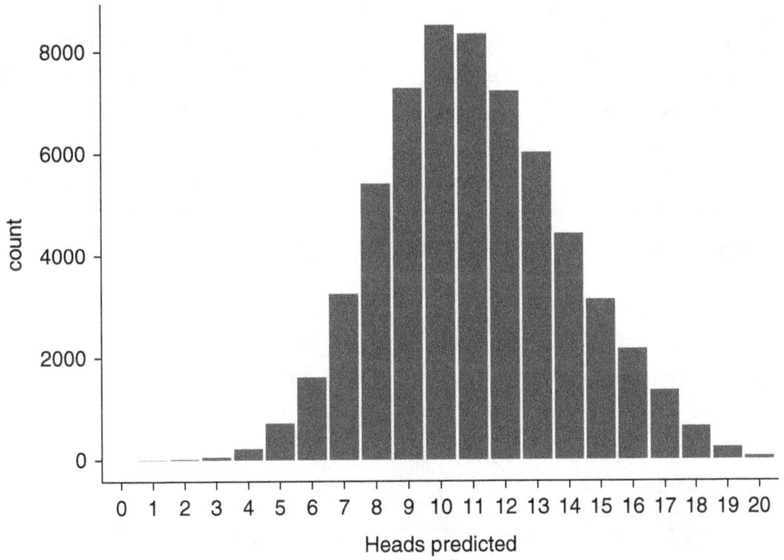

Figure 6.4 Prediction of number of heads in 20 coin flips, based on previous analysis of possible θ value

The number of heads in 20 'new' flips is predicted in Figure 6.4. Note that the number of heads is converted into a factor. This is in case some number of heads does not appear in the experiment (e.g. a very unlikely outcome such as 19 heads). It forces geom_bar to add this outcome to the axis since it is one of the possible *levels* of the factor.

This is a realistic representation – since the coin flipping is a random process, an estimate of a single number without further explanation would be misleading. Instead, we have a range of possible outcomes and their relative likelihood. If a single number is really needed, the mean of this sample could be calculated:

```
# define a factor to integer conversion function
factor_to_int <- function(x) as.integer(as.character(x))
predictions %>%
   mutate(`Heads predicted`=factor_to_int(`Heads predicted`))   %>%
   summarise(`E(Heads predicted)`=mean(`Heads predicted`))
## # A tibble: 1 x 1
##    `E(Heads predicted)`
##                   <dbl>
## 1                  11.1
```

In this calculation we need to turn the factor-based Heads predicted variable back into an integer. Unfortunately, direct conversion via as.integer just returns a numeric code for each level, from 1 to the number of levels. This means a predicted outcome of 0 heads gets the code 1, and so on. However, by converting to character first, we get the *label* for each level as a character variable. This is a character representation of the integer amount actually wanted – and can be turned into a 'genuine' integer via as.integer – this is what factor_to_int does.

Thus a point estimate gives a value of just over 11 as the expected number of heads in 20 flips (this is influenced by the relatively high number of heads flipped) and the likelihood that if the coin is a fake then θ is above 0.5.

A compromise to someone requiring a number as an answer to their prediction question might be to identify upper and lower limits – for example, upper and lower 2.5% quantiles of the distribution. This is just two numbers, but does convey the message that the outcome cannot be predicted deterministically:

```
predictions %>%
   mutate(`Heads predicted`=factor_to_int(`Heads predicted`)) %>%
   summarise(`Heads predicted - 2.5%`=quantile(`Heads
predicted`,0.025),
             `Heads predicted -97.5%`=quantile(`Heads
predicted`,0.975))
## # A tibble: 1 x 2
##    `Heads predicted - 2.5%` `Heads predicted -97.5%`
##                      <dbl>                    <dbl>
## 1                        6                        7
```

So the outcome is still quite uncertain. This is something we will term the *paradox of replicability*. Even if a very large number of observations have been made of a random process, this makes the process no less random. All the large number of observations provide us with – providing they are appropriate observations – is a very reliable picture of the nature of the randomness. However, some seem to believe that more data will give less error in predictions.

Here, this can be illustrated by imagining that instead of 20 coin flips (and 13 heads) you had 2000 coin flips with 1300 heads. This would give a much more certain estimate of θ with almost no evidence supporting the fact that the genuine coin was chosen:

```
set.seed(6371)
# Because there are many more possible outcomes we need to increase
# the number of flips to get a reasonable sample with exactly 1300
# heads
n_sims <- 100000000
## Then things are done pretty much as before
theta <- ifelse(runif(n_sims) < 0.5, runif(n_sims), 0.5)
results4 <- tibble(Theta=theta, Heads=rbinom(n_sims, 2000, theta))
results4 %>%
  filter(Heads==1300) %>%
  mutate(Hp5=Theta==0.5) %>%
  count(Hp5) %>%
  transmute(Hp5, prob=n/sum(n))

## # A tibble: 1 x 2
##     Hp5      prob
##   <lgl>    <dbl>
## 1 FALSE        1
```

This shows that there are no simulations that result in *exactly* 1300 heads out of 2000 that gave values of $\theta = 0.500$. However, these results can then be used to create a predictive distribution for the number of heads in 2000 flips:

```
set.seed(5780)  # Reproducibility!
# Simulate the future coin flipping
results4 %>%
  filter(Heads==1300) %>%
  mutate(`Heads
predicted`=factor(rbinom(nrow(.), 20, Theta), levels=0:20)) ->
predictions2
# Show predicted outcome as a histogram
predictions2 %>%
  ggplot(aes(x=`Heads predicted`)) +
```

```
geom_bar(fill='firebrick') +
scale_x_discrete(drop=FALSE)
```

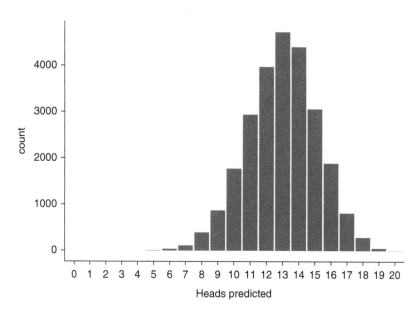

Figure 6.5 Predictive distribution based on the 2000 coin flip experiment, for 20 new flips

The results in Figure 6.5 suggest there is now much stronger evidence that the probability of flipping a head is 13/20 or 0.65, so that 13 heads is a much more likely outcome than was suggested on the basis of just 20 flips – but there is still plenty of uncertainty!

This last experiment also suggests another characteristic of the kind of experimental prediction used here. It is possible to predict the outcome of processes that were not identical to the one used to estimate the unknown parameter. In that last example the difference was trivial – using a 2000 coin flip experiment to predict a 20 coin flip process. But providing appropriate characteristics carry through, it is possible to predict outcomes of relatively different processes. Thus if the same coin were used in a different experiment, outcomes could be predicted using the experimental distribution for θ already obtained.

A more subtle example will be illustrated next. Suppose you have the evidence based on the 20 coin flips (i.e. 13 heads, simulated possible θ values stored in `results3`) but in a new experiment the coin will be flipped until six heads are seen. Here you wish to predict how many coin flips this is likely to take. That can be modelled using the *negative binomial distribution*. If n_t is the number of tails thrown in the sequence of flips carried out until n_h heads are thrown, then

$$P\left(n_t\right) = \binom{n_t + n_h - 1}{n_t} \theta^{n_h}(1-\theta)^{n_t}$$

(6.12)

Thus, the total number of flips is just $n_t + n_h$ where the distribution for n_t is given by equation (6.12), and n_h is fixed (here it takes the value 6). In R, the function rnbinom generates random numbers from this distribution. Thus an experimental prediction for the total number of coin tosses (based on the 13 out of 20 outcome) can be created as follows:

```
set.seed(12011)
# define an integer to factor conversion function
int_to_factor <- function(x) factor(x,levels=min(x):max(x))
results3 %>%
   filter(Heads==13) %>%
   mutate(`Heads predicted` =
             int_to_factor(rnbinom(nrow(.),size=6,prob=Theta)+6)) ->
predictions
predictions %>%
   ggplot(aes(x=`Heads predicted`)) +
   geom_bar(fill='firebrick') + scale_x_discrete(drop=FALSE)
```

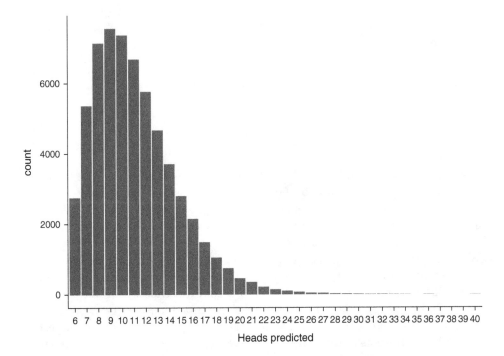

Figure 6.6   Predicted distribution of the number of flips need to get six heads

Again, the outcome is not completely predictable, but it seems more likely that around 8–10 flips will be needed to obtain the six heads (Figure 6.6).

Key Points

- Models are used to answer questions with some indication of the reliability of the answer.

- One approach to reliability/uncertainty is through probability, which indicates the likelihoods of different outcomes to the question.

- Simulated models of these theoretical processes are relatively simple to encode in R.

- If the outcomes are unknown then these can be simulated with suitable parameterisation in order to compute experimental probabilities of different answers (such as the coin being true or fake).

- It is easy to generate likelihoods for different theoretical questions and answers (e.g. if the probability of a fake changes to 0.01, or if the question changes to the probability of getting a head).

- Often there is no answer to true/false statements, rather there are probabilities of a value being within a range (e.g. the difficulty in determining the probability that $\theta = 0.5$ *exactly* versus being between 0.49 and 0.51).

- This is done by examining the probability distributions (θ) that lead to the observed outcome (e.g. of 13 heads out of 20).

6.3 MORE CONCEPTUALLY DEMANDING QUESTIONS

The above section could be said to make use of simplistic examples, but the motivation was to introduce some of the typical validation and assessment questions asked when modelling and to set out one possible experimental methodology for answering them. The process being modelled was that of flipping a coin, and although this was modelling a random process, in one sense it resembles van der Waals' approach as set out in Section 6.1: for each of the processes described there is a clearly defined (and derivable) model for the outcome based on a set of assumptions about the coin flips, the way the coin was chosen, and the characteristics of fake and genuine coins. However, as described in Section 6.1, this kind of rigorous derivation may not be possible. For example, it may be of interest to model house price as a function of local unemployment rates. Whereas it seems plausible that the two variables may be connected, there is no immediately obvious functional relationship between the two quantities. This section provides a worked example in a more 'real-world' context of assessing the validity of two different potential house price models.

You will need to clear your workspace, load some packages and then download and load the ch6.RData file to your R/RStudio session:

```
# load packages
library(tidyverse)
library(dbplyr)
library(sf)
library(tmap)
library(RColorBrewer)
library(spdep)
```

The code below downloads data for the worked example to your current working directory. You may want to check this:

```
# get the current working directory
getwd()
# clear the workspace
rm(list = ls())
# download data
download.file("http://archive.researchdata.leeds.ac.uk/740/1/
              ch6.Rdata",
              "./ch6.RData", mode = "wb")
# load the data
load("ch6.RData")
# examine
ls()
## [1] "lsoa"        "oa"           "properties"
```

This loads three spatial datasets in sf format as follows:

- properties, a multipoint object of houses for sale in the Liverpool area;

- oa, a multipolygon object of Liverpool census Output Areas (OAs) with some census attributes;

- lsoa, a multipolygon object of Liverpool census Lower Super Output Areas (LSOAs) with an LSOA code attribute.

The OAs nest into the LSOAs (typically each LSOA contains around five OAs). The LSOAs will be used in this section to construct and evaluate the models to be compared. The datasets, their spatial frameworks and their attributes are fully described in Chapter 7, where they are explored and used in much greater depth.

6.3.1 House price problem

Consider the task of constructing different house price models using socio-economic data and deciding which one to use. This decision should be underpinned by some form of model validation and assessment.

Two house price models are assessed in terms of their ability to reproduce characteristics of the input dataset. Two ideas are useful, and after they are explored, the model evaluation ideas from the coin-tossing experiment can be applied to assess the two models. The two ideas are as follows:

- A model is of practical use if it can simulate datasets that are similar in character to the actual data.

- However, different models can be useful in different ways and in different situations we may define 'similar in character' differently.

The two house price models both consider house prices to be a function of geographical, social and physical factors, but with the factors provided in different ways. The physical factors are:

- whether the house is terraced;

- whether any bedrooms have en suite facilities;

- whether it has a garage;

- the number of bedrooms it has.

The geographical factor is simply the northing value of the property. The data used in the analysis (compiled below) are projected in the OSGB projection, which records location in metres as an easting (latitude) and a northing (longitude). The northing is the y coordinate of the house's location and in the Liverpool area (Merseyside) there is a notable north/south trend in house price.

The two models model the social factors differently. For model 1, a number of variables representing social and demographic conditions of the local OA are included:

- unemployment rate;

- proportion of population under 65;

- proportion of green space in the area (a crude proxy for quality of life).

This is a relatively limited number of variables, but crudely encapsulates socio-economic conditions, age structure and quality of life. However, these are, at best, blunt instruments, although all of these factors correlate with house prices.

Model 2 takes a different approach to representing the socio-economic conditions of a house's locality. In this case, the model uses a geo-demographic classification of each census OA, as defined by the OAC as introduced in earlier chapters. Each OA is classified into one of eight groups, seven of which are present in the Liverpool data:

- constrained city dwellers;
- cosmopolitans;
- countryside (not present in Liverpool!);
- ethnicity central;
- hard-pressed living;
- multicultural metropolitans;
- suburbanites;
- urbanites.

The OAC, a categorical variable, is used as a predictor. Unlike the approach in model 1, the OAC groups are the distillation of a large number of variables (see Vickers and Rees, 2007; Gale et al., 2016). However, by reducing this information into just eight distinct classes, arguably some information is lost. Here, the interest is in which of these two approaches – both with strengths and weaknesses – best models the property price data.

6.3.2 The underlying method

In more formal terms, both models will model log house price as a linear function of the predictor variables. Thus, model 1 in R modelling notation is defined as follows:

```
log(Price)~unmplyd+Terraced+gs_area+Ensuite+Garage+as.numeric(Beds)+
Northing+u65
```

Model 2 is similarly defined as follows:

```
log(Price)~OAC+Terraced+Ensuite+Garage+as.numeric(Beds)+Northing
```

Both essentially take the form

$$\log(y_i) = \beta_0 + \sum_{j=1}^{k_j} \beta_j x_{ij} + \epsilon_i$$

(6.13)

where

$$\epsilon_i \sim N\left(0, \sigma_j^2\right)$$

(6.14)

in which σ_j^2 is the variance for model y and k_j is the number of explanatory variables in model j, for $j = \{1, 2\}$.

In each case, we can simulate the price of each house by estimating the variance of the regression error term σ_j^2 (basically the mean residual sum of squares), perturbing the predicted log price by a random normal variable with mean zero and variance σ_j^2, and then taking the antilogarithm (the exp function).

Thus we can simulate property price datasets for each of the models. The question is then how similar the simulated datasets are to the actual one. Reverting back to the earlier points, we must first decide in which way *similarity* is defined.

Here, suppose geographical patterning is of interest. One way we could measure this is by spatial autocorrelation. This could be done by computing the mean house price for each LSOA in Merseyside, and computing Moran's I coefficient, a measure of spatial autocorrelation (Cliff and Ord, 1981), for the LSOA. The value of these coefficients for several simulations of each model could be compared with the value for the observed data constructed using OAs.

At this point, the coin-tossing idea comes into play. Starting with simulating datasets from models 1 and 2 in equal amounts (e.g. 1000 from model 1 and 1000 from model 2), ones *not* similar to the actual data (in terms of Moran's I as described above) are discarded, so that only those that are similar to the actual data are retained.

The relative proportions of datasets generated by the two models lead to an approximate estimate of the probability that each model is the 'correct' one, given that one of the models is correct.

6.3.3 Practical computation in R

The data were loaded earlier and some data manipulation is needed to format the data for modelling and simulation. This is a lengthy series of transformations – and typical of a much 'real-life' data science, where much effort is directed to reformatting and cleaning data rather than the actual analytical effort.

The code block below does the following to create data_anal:

1. Transforms the properties data from WGS84 coordinates to the OSGB projection (EPSG code 27700) and extracts the easting and northing of each property address, adding it to the data frame.

2. Assigns each property to its enclosing census OA to link OAC groups and socio-economic data from the 2011 UK Census.

3. Assigns each property to an enclosing LSOA to compute Moran's I.

4. Removes the geometry from the object (to fit models as a simple data frame).

5. Removes records with missing values and selects the desired variables.

```
data_anal <- properties %>%
    # 1. transform and calculate Easting and Northing
    st_transform(27700) %>%
    mutate(Easting = (properties %>% st_coordinates() %>% .[,1])) %>%
    mutate(Northing = (properties %>% st_coordinates() %>% .[,2])) %>%
    # 2. join with OA data to attach socio-economic data
    st_join(oa) %>%
    # 3. join to LSOA to attach the LSOA code
    select(-code) %>%
    st_join(lsoa %>% select(code)) %>%
    # 4. remove the sf geometry
    st_drop_geometry() %>%
    # 5. remove any records with NA values, select desired variables
    drop_na() %>%
    select(Price,OAC,Northing,Easting,gs_area,code,unmplyd,
           Terraced,'En suite',Garage,Beds,u65,gs_area) %>%
    mutate(code=as.character(code))
```

The two models can now be fitted:

```
model1 <- lm(log(Price)~unmplyd+Terraced+`En suite`+Garage+as.
numeric(Beds)+
                Northing+u65+gs_area, data = data_anal)

model2 <- lm(log(Price)~Terraced+`En
suite`+Garage+as.numeric(Beds)+OAC+
                Northing, data = data_anal)
```

They may then be inspected:

```
summary(model1)
summary(model2)
```

In both cases, coefficients seem to take the sign expected – for example, the presence of a garage increases value.

We may now create simulated datasets for both of the models. The function get_s2 estimates σ_j^2 (the variance) and the function simulate creates the simulation of the data – in terms of simulated house prices for each property. Here 1000 simulations of each dataset are created and stored in the matrices m1_sims and m2_sims. Each column in these matrices corresponds to one simulated set of prices for each property.

```
# Helper functions - here
# 'm' refers to a model
# 'p' a prediction
# 's' a simulated set of prices
get_s2 <- function(m) sqrt(sum(m$residuals^2)/m$df.residual)
simulate <- function(m) {
  # get the model predictions of price
  p <- predict(m)
  # a random distribution with sd equal to the variance
  r <- rnorm(length(p),mean=0,sd=get_s2(m))
  # create a simulated set of prices and return the result
  s <- exp(p + r)
  return(s)}
# set up results matrices
m1_sims <- matrix(0,length(data_anal$Price),1000)
m2_sims <- matrix(0,length(data_anal$Price),1000)
# set a random seed and run the simulations
set.seed(19800518)
for (i in 1:1000) {
  m1_sims[,i] <- simulate(model1)
  m2_sims[,i] <- simulate(model2)
}
```

Next, the computation of Moran's *I* of the mean house price for properties in each LSOA is undertaken using tools in spdep and aggregation via group_by. Note that as not all LSOAs have a property in them, the simple features object lsoa_mp has a small number of 'holes' in it:

```
# determine LSOA mean house price
lsoa_mp <- data_anal %>%
  group_by(code) %>%
  summarise(mp=mean(Price)) %>%
  left_join(lsoa,.) %>%
  filter(!is.na(mp))
# create a weighted neighbour list of the LSOA areas
lw <- nb2listw(poly2nb(lsoa_mp))
# determine the spatial autocorrelation of LSOA mean house price
target <- moran.test(lsoa_mp$mp,lw)$estimate[1]
```

Looking at the value of target (so named because we want the simulated values to be close to this value of Moran's *I*), we see that it is fairly high at around 0.61:

```
# simulate Moran's I value
sim_moran <- function(m_sims,lw) {
  res <- NULL
```

```
for (i in 1:ncol(m_sims)) {
  data_sim <- data_anal %>% mutate(Price=m_sims[,i])
  lsoa_mp <- data_sim %>%
    group_by(code) %>%
    summarise(mp=mean(Price)) %>%
    left_join(lsoa,.,by=c('code'='code')) %>%
    filter(!is.na(mp))
  res <- c(res,moran.test(lsoa_mp$mp,lw)$estimate[1])
}
return(res)
}
# evaluate both models
test1 <- sim_moran(m1_sims,lw)
test2 <- sim_moran(m2_sims,lw)
```

Now, test1 and test2 contain lists of the simulated values of Moran's *I* for each model. The assessment of the two models is carried out by seeing how many of each sample are sufficiently close to the observed value. An exact match is more or less impossible, but we could, for example, say that a Moran's *I* within 0.05 is acceptable. Thus, the count of such acceptable values for each group is computed for model 1 (t1) and model 2 (t2):

```
t1 <- sum(abs(test1 - target) < 0.05)
t2 <- sum(abs(test2 - target) < 0.05)
c(t1,t2)

## [1] 554 253
```

This suggests that the probability of model 1 being acceptable compared with model 2 is 0.686. This suggests that the odds based on this principle are around 2 : 1 in favour of model 1. A few points need to be considered here:

- This suggests that a model using small number of census-based variables to represent the social and demographic aspects of a locality better reflects spatial clustering in house prices than one using a geo-demographic classification.

- However, this is a very specific criterion. It is possible that in other situations the other model may be more appropriate.

- Here there were just two 'contender' models, but an alternative may outperform both of these.

A further point to consider is that although no model reflects reality perfectly, some models reflect certain aspects of reality well. The exercise above gives a practical approach to deciding how well a set of alternative models reflects some aspect of reality.

Key Points

- The rigorous derivation of probabilities and likelihoods of outcomes may not always be possible.

- The worked example demonstrates how to determine which of two competing house price models to use, underpinned by model validation and assessment.

- For each approach a regression model of log price was created.

- The house price data were simulated, in this case by estimating the variance of the regression error term and perturbing this by including a random term in the predicted log price, with a fixed distribution.

- This allows the simulated data to be compared with the real price data to evaluate the models.

- However, measures of 'similarity' need to be determined. In the house price example, this was done by simulating values of Moran's I for each model and then determining how many times the simulated data were sufficiently close to the observed value.

6.4 MORE TECHNICALLY DEMANDING QUESTIONS

Another issue arises where, even if the question is simply posed, there are technical, computational challenges to answering it. Examples where this situation arises include scenarios where:

1. datasets are very large, and computationally it is demanding to apply approaches which may be usable for smaller datasets;

2. theoretical distributions are used, but are difficult to model or simulate;

3. theoretical distributions are used, but it is difficult (or impossible) to find an algebraically expressed likelihood function, as required in a Bayesian or classical approach to inferential statistics.

6.4.1 An example: fitting generalised linear models

This is an example of situation 1 above. Generalised linear models (GLMs) are powerful statistical tools allowing the regression of count data, binary data and continuous data onto categorical and continuous predictors. Fitting a GLM is well understood computationally, although for large datasets the computational time is large. To recap, a GLM takes the form

$$g\left(E(y_i)\right) = \sum_{j=1,\ldots,m} \beta_j x_{ij} \qquad (6.15)$$

where $g(.)$ is a link function, and the distribution of y_i is a member of the *exponential family* of distributions. This includes, for example, the Gaussian, binomial and Poisson distributions. The link function can take many forms, but typically it transforms the domain of $E(y_i)$ onto the range $(-\infty, \infty)$, so that it maps onto the possible values of the linear predictor on the right-hand side. For example, in the Poisson case $E(y_i)$ is always positive, and a common link function is log since it maps positive values onto $(-\infty, \infty)$. If $\mathbf{g}(.)$ denotes the vectorised version of $g(.)$, then the model can be written in vector–matrix notation as

$$\mathbf{g}(E(\mathbf{y})) = \mathbf{X} \qquad (6.16)$$

Typically this is a useful tool for predicting y or identifying the effect of a number of predictor variables on y. All of this is made possible by estimating the β_i values given the data $\{y_i, x_i, \forall i, j\}$. Unless the distribution of y_i is Gaussian, the estimation is an iterative process – see, for example, McCullagh and Nelder (1989). Of computational note – and of relevance here – is that the iterative process works by starting with an initial guess for the values of $\{\beta_1,\ldots, \beta_{im}\}$ and successively updating this until subsequent estimates are sufficiently similar. This involves fitting weighted ordinary least squares (OLS) regression models whose weights and dependent variable depend on the current coefficient vales.

However, as noted above, although this is a well-known technique, for large datasets the computation time can be prohibitive, as can the memory requirements. Examining the process in more detail, the calculations described involve iterative OLS regression calibrations. If $\mathbf{X}$ is the design matrix (i.e. the matrix of predictor variables), and $\mathbf{y}$ is the vector of the dependent variable, then the coefficients β are estimated by solving

$$(\mathbf{X}^T \mathbf{X})\beta = \mathbf{X}^T \mathbf{y} \qquad (6.17)$$

This can be achieved by *Gaussian elimination* – see, for example, Atkinson (1989). If there are a large number (m) of prediction variables then this can be time-consuming – the matrix equation above is actually a system of m linear equations. However, when it is n, the number of observations, that is large, the issue is not the solution of the equations, but the construction of the matrices $\mathbf{X}^T\mathbf{X}$ and $\mathbf{X}^T\mathbf{y}$. Each element of these matrices is the sum of products of the elements in the columns in $\mathbf{X}$ and $\mathbf{y}$. The (i, j)th element of $\mathbf{X}^T\mathbf{X}$ is $\sum_{k=1,\ldots,n} x_{ik} x_{kj}$ and the (i, j)th column of $\mathbf{X}^T\mathbf{y}$ is $\sum_{k=1,\ldots,n} x_{ik} y_k$. If n is very large (e.g. several million observations) then it is these cross-product summations that may be time-consuming. In addition, storage of **X** could lead to problems, if it is too large to fit into memory.

These considerations are unlikely to impinge on an 'ordinary' regression analysis, but for one carried out on a large dataset these are complications to be addressed. Typically, for the large-n scenario, one solution may be to carry out as much of the analysis as possible within an external database (e.g. via a linked SQLite object). Also, sums and products may be dealt with via SQL commands, although the Gaussian elimination generally needs to be dealt with directly in R.

Another way that the summation of cross-products can be achieved is by reading in 'chunks' of the **X** and **y** matrices from a database into memory, and computing the cross-product sums incrementally. This way, the entire dataset is never read directly into R. This is the approach used in the package biglm, at least when fitting a simple OLS model.

This is illustrated below. Here a regression model is applied to the prescription data described earlier in this book, particularly for prescriptions of drugs with the BNF code of '0403' (essentially antidepressants). On this occasion the regression model is applied not to data aggregated over OAs or LSOAs, but at the individual prescription level. Here the actual prescription cost will be regressed against the education and employment components of the LSOA data introduced in Chapter 4.

First, a number of packages are needed to do this:

- dbplyr and dplyr to manipulate data in an external database (recall that dplyr has a dependency on dbplyr);

- biglm to fit regression models to external data;

- DBI and RSQLite to access the external prescription database.

```
library(dbplyr)
library(dplyr)
library(biglm)
library(DBI)
library(RSQLite)
```

Next, a connection is set up. Note that you may have to change your working directory to the folder location of the 11.2 GB file you downloaded in Chapter 4:

```
db <- dbConnect(SQLite(), dbname="prescribing.sqlite")
```

Once this is done, the database is accessed – in particular, the prescriptions table is joined to the practices table (via the key practice_id) which is then joined to the table postcodes, which provides a link via the lsoa_id to the social table. Thus, for each prescription, associated socio-economic variables (linked to the practice address) are available. The code below creates a query that generates a table as a potential input to a 'big' linear regression model:

```
cost_tab <-
  tbl(db, "prescriptions") %>%
  # get the antidepressants
  filter(bnf_code %like% '0403%') %>%
  # join to practices (this will be done by practice_id)
  left_join(tbl(db, "practices")) %>%
  # join to postcodes (this will be done by postcode)
  left_join(tbl(db, "postcodes")) %>%
  select(act_cost, postcode, lsoa_id) %>%
  left_join(tbl(db,"social"))
```

Recall that, working with dbplyr, the tbl command here does not immediately create a table in R; it simply associates a database query to extract a table from a database connection and stores this in an object. All of the various dplyr-like operations essentially modify this query, so that cost_tab associates the resultant (quite complex) SQL query to obtain the table created at the end of this process.

So although it looks like a table, it is in fact an encoded query to a database, with some tibble-like properties. You can check this by examining the class of cost_tab:

```
class(cost_tab)
```

The query that is encapsulated by cost_tab can be extracted via the sql_render function in the dbplyr package. This works best if the result (an sql object) is converted to character and printed via cat; the result is then more 'human readable', although still quite complicated:

```
sql_render(cost_tab) %>% as.character() %>% cat()
```

If the above query were sent to the database connection db this would return the dataset used for the planned regression. However, there are around 3.7 million records here (the large-n problem described above). By using dbSendQuery and dbFetch it is possible to retrieve the results of the query in chunks. Here chunks of 400,000 observations are used. In the code below, first a query is sent to the database connection (using dbSendQuery). The result of the query is not a data frame itself, but another kind of connection (rather like a file handle) that can be used to fetch the data – either as a whole or in chunks of n observations. Here, the chunking approach is used. The handle (here stored in cost_res) bookmarks its location in the table generated by the query, so that successive calls lead to successive chunks of the resultant data frame being returned.

In turn, biglm works in a similar way. The first call fits a linear regression model to the first chunk of data. The following calls to update will update the model, giving the result of a linear regression model, and add an extra chunk of data added to the model. This works using the 'chunked' approach to the forming of (X^TX) and

X^Ty – and also keeping a running estimate of β. Thus, the approach to fitting a large regression model is to read a chunk and add a chunk to the model, until all of the dataset has been added. When the entire query has been fetched, an empty database is returned. Thus, the approach described may be set up as a loop, with the exit condition being an empty data frame returned from dbFetch. Finally, the dbClearResult closes off the handle (similar to closing a file), freeing up the database connection for any future queries, and the database is closed:

```
# Obtain the query used to obtain the required data
cost_query <- sql_render(cost_tab) %>% as.character()
# Set up the query to the database connection db
cost_res <- dbSendQuery(db, cost_query)
# Fetch the first chunk
chunk <- dbFetch(cost_res, n=400000)
cost_lm <- biglm(act_cost~ unemployed + noqual, data=chunk)
# Keep fetching chunks and updating the model until none are
left
counter = 1 # a counter to help with progress
repeat{
    chunk <- dbFetch(cost_res, n=400000)
    if (nrow(chunk) == 0) break
    cost_lm <- update(cost_lm, chunk)
    cat(counter, "\t")
    counter = counter+1
}
# Close down the query
dbClearResult(cost_res)
```

Once this step has been completed, the model may be inspected, to see the fitted coefficient values, or predict new values:

```
# Print the results
summary(cost_lm)
```

Here the fitted coefficients (and their standard errors) show that coefficients for low levels of education (noqual) have a positive linkage with prescription expenditure on antidepressants, and perhaps surprisingly unemployment is negatively associated.

This demonstrates how OLS regression can be applied to large datasets. However, earlier *generalised* linear models were mentioned and illustrated by equation (6.15). Recall that they are calibrated using an iteratively refitted regression model. Thus, a process such as that above has to be applied iteratively. This would require some coding to be set up, including a 'first principles' implementation of the GLM calibration procedure with the 'chunking' OLS method embedded. Fortunately, this is simplified with the big GLM function bigglm in the biglm library.

This works in a similar way to `biglm` but is capable of fitting models where the y-variable has distributions other than Gaussian. For example, it could be assumed that y has a gamma distribution, and that g is the log function. The distribution of the actual prescription cost is then modelled as

$$f(y) = \frac{v}{\mu\Gamma(v)}\left(\frac{vy}{\mu}\right)^{v-1} e^{-\frac{v}{\mu}y} \tag{6.18}$$

where μ is the expected value of y. v is a shape parameter: when $v = 1$, y has an exponential distribution, and when $v > 1$ it has an asymmetric distribution with a long tail (see Figure 6.7). Here, $E(y) = \mu$ and $Var(y) = y^{-1}$.

This model seems sensible. Although it cannot be justified in the rigorous way that van der Waals' model can be, it provides a reasonable schema that may justify exploration. Here we are considering the situation where prescription costs depend on some aspects of socio-economic deprivation, and where they also have a distribution with a relatively heavy 'upper tail' where a small number of prescriptions are very expensive. Also, the logarithmic link suggests that regression effects are multiplicative, so that for more expensive drugs, a unit change in a deprivation-related attribute (qualifications, unemployment, etc.) is associated with a change in prescription that is *proportional* to the cost of the drug.

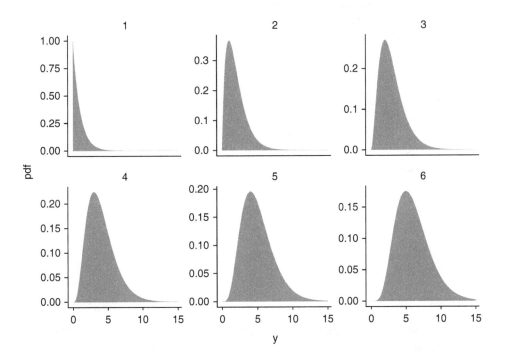

Figure 6.7　The gamma distribution for different values of v

6.4.2 Practical considerations

The justification for a log-gamma model has been set up above, but the practicalities of fitting such a model to a dataset with around 3.7 million records must now be considered. As observed, this is a GLM, and requires iterative refitting, as is done if bigglm is used. This works in a similar way to the standard glm package. A key difference is that instead of supplying a data frame, a *function* is supplied that returns data frames for successive chunks of the data used each time it is called. A further requirement of this function is that it contains just one argument, named reset. This is a logical argument: if TRUE then the function resets the data-fetching process to begin at the first observation in the database query; if FALSE then the next chunk is returned. One useful way to handle functions of this kind is to create another more generic function to return a function for this specific database and query. The function make_fetcher below does this. This function takes a database connection, an SQL query and a chunksize saying how many records are to be fetched each time the returned function is called. Note that the res variable is created inside make_fetcher but not inside the returned function. However, the returned function does refer to this, and even assigns something to it in when reset is TRUE. This is achieved via the <<- operator, which assigns to variables in the surrounding environments, rather than inside the function itself. This effectively makes the res variable *persistent*, so that its status from the previous function call is preserved at the start of the following call – and this achieves the requirement that the function call returns the *next* chunk of the query result each time it is called:

```
make_fetcher <-function(db, query, chunksize,...){
    res <- NULL
    function(reset=FALSE){
    if(reset){
       res <<- dbSendQuery(db,query)
    } else{
       rval <-fetch(res,chunksize)
       if (nrow(rval)==0) {
            rval<-NULL
            dbClearResult(res)
       }
       return(rval)
    }
    }
}
```

Once this function – essentially a tool to set up the bigglm call – has been defined, the model may be fitted. The query used to extract the data needed is still stored in cost_query, so the fetching function may be made based on this. Similarly, the database connection was set up earlier, and stored in **db**. Thus, the model may be set up as below, and this will take time to run – times will vary depending on your computer!

```
library(lubridate) # To time the process
# Call the 'fetcher' function
cost_df <- make_fetcher(db,cost_query,500000)
# Run the glm
starting_point <- now()
cost_glm <- bigglm(act_cost~ unemployed + noqual,
                   data=cost_df,
                   family=Gamma(log),
                   start=c(4.2,-0.004,1.007),
                   maxit=35)
duration <- now() - starting_point
duration

## Time difference of 4.153473 mins
```

Note that the **maxit** parameter (controlling the maximum number of iterations that may be applied) is set to quite a high value (35). For large datasets, the authors have observed that iterative calibration of GLMs can converge quite slowly. This was the case here, hence the choice of parameter. Similarly, a good initial guess for parameters tends to give better results for convergence. Here, the values were chosen by selecting a subset of the data and estimating the parameters based on the subset.

The results of the model may be seen below:

```
summary(cost_glm)$mat
##                  Coef      (95%       CI)          SE p
## (Intercept)  3.835381  3.829197  3.841565 0.003091956 0
## unemployed  -2.523984 -2.591434 -2.456535 0.033724710 0
## noqual       1.114739  1.082907  1.146571 0.015915960 0
```

Note that the *p*-values associated with the coefficients are very low, suggesting that there is strong evidence that they are not zero. However, it should also be noted that the value of **unemployed** is very low, which suggests that although not zero, the effect attributable to it is very small – a unit change in the unemployment score results in a change in the log of the expected prediction cost of –2.5240, or a multiplicative factor of around 0.0801, suggesting minimal practical implications. However, the **noqual** factor, at 1.1147 for the logged expectancy, translates to a multiplicative change of 3.0488, a more notable effect, with a multiplicative increase in costs of around 3 for each antidepressant prescription (suggesting larger doses?) for each additional percentage of the population with no qualifications.

6.4.3 A random subset for regressions

Above it was noted that a good starting guess for the regression coefficients was helpful, and that this could be obtained by initially calibrating the model with a subset of the full data, chosen at random. One issue here is that selecting a random subset from an external database also requires consideration. Such a subset can be selected via an SQL query, but sometimes there are some issues. First, although dbplyr can be used on an external database, operations like mutate are restricted to the operations permitted by the SQL interface of the external database. An issue here is that random number generation in SQLite is far less flexible than that provided in R, essentially consisting of a single function random that generates random 64-bit signed integers. These are integers between −9,223,372,036,854,775,808 and +9,223,372,036,854,775,807 with equal probabilities of occurrence. There are two ways that random could be used to select a random subset, of size M, say:

1. Assign a random number to each record, sort the data according to this number, then select the first M of them.

2. By using modulo and division, alter the range of the random number to [0, 1] and then select those records where the new number is below M/N. This can be achieved approximately by rescaling to [0, 1,000,000] and making the threshold 1,000,000M/N.

Option 1 has the advantage of guaranteeing exactly M cases in the subset, whereas option 2 provides this on average. However, option 1 requires sorting a large external database (which can take some time), whereas option 2 does not. Since in this situation the *exact* size of the subset is not particularly important, option 2 is used. This can be done using dbplyr but extending the original query used to extract the required prescription data. Suppose here a subsample of size 4443 is required.

This can be achieved with the code below. First the random subset selection is applied to the cost_tab external table option. This is then fed to the standard glm function (not bigglm, since the dataset is smaller). The collect function here takes the remote connection database query and converts it to a standard tibble:

```
# Augment the dataquery
set.seed(290162)
cost_glm_sample <- cost_tab %>%
    mutate(r = abs(random() %% 1000000)) %>%
    filter(r < 1000000*4443.0/3678372) %>% collect()
ss_cost_glm <- glm(act_cost~ unemployed + noqual,
                   data=cost_glm_sample,
                   family=Gamma(log))
coef(ss_cost_glm)
## (Intercept)   unemployed        noqual
##    3.814116    -2.251489      1.088021
```

Your results will be slightly different. To ensure reproducibility, the same random seed *should* be used each time. However, the SQLite random() function does not allow a seed to be specified explicitly, despite its being set in the code snippet above. Alternative approaches (possibly using different database drivers or creating the random numbers in R and writing them to the database) will be necessary if reproducibility is required. One workaround is to save the randomly selected subset as an extra table in the external database connection – although the act of generating the subset cannot be made reproducible, the actual subset thus generated may be preserved. The following code saves the dataset in glm_ prescription_sample:

```
dbWriteTable(db,"glm_prescription_sample",cost_glm_sample)
```

This is included in the SQLite database file prescribing.sqlite – thus the same random subsample can now be reused.

6.4.4 Speeding up the GLM estimation

There is one further approach to speeding up GLM parameter estimation for large datasets, using a suggestion by Lumley (2018). In this article it is noted that just a single iteration of the GLM regression algorithm from a sample of size N is reasonably close to the converged result when the initial 'guess' is the maximum likelihood estimate drawn from a sample of size $N^{1/2+\delta}$, where $\delta \in (0, 1/2]$. In short, provided a random subsample of a size slightly larger than $\sqrt{N}$ is used to calibrate the initial guess, a single iteration of the GLM algorithm provides a workable estimate of the regression coefficients. Here, a subsample of size just over $N^{5/9}$ was used – hence the choice of subsample size in the example.

Using the set of estimates obtained from the random subsample but restricting the bigglm algorithm to a single iteration gives the results below:

```
# Call the 'fetcher' function
cost_df <- make_fetcher(db,cost_query,500000)
# Timing
start_point_1it <- now()
# Run the glm
cost_glm_1it <- bigglm(act_cost~ unemployed + noqual,
                  data=cost_df,
                  family=Gamma(log),
                  start=c(4.2,-0.004,1.007),
                  maxit=1)
duration_1it <- now() - start_point_1it
summary(cost_glm_1it)$mat
```

```
##                     Coef      (95%      CI)          SE p
## (Intercept)     3.885269  3.881564  3.888975 0.001852770 0
## unemployed     -1.366195 -1.406613 -1.325778 0.020208609 0
## noqual          1.049599  1.030525  1.068674 0.009537204 0
```

In this instance, ignore the warning about failure to converge – this is simply due to the setting of maxit to 1. As can be seen in the summary output, the results are quite similar to those when iterating until convergence. Here the time taken is notably shorter, again allowing for differences between computers:

```
duration_1it
## Time difference of 26.35219 secs
```

As can be seen, although the results are slightly less accurate, the model fitting has been speeded up by a factor of around 7.9.

Finally, close the database:

```
dbDisconnect(db)
```

Key Points

- Generalised linear models are powerful statistical tools for working with data whose distribution is a member of the *exponential family* of distributions.

- They require a link function to transform the target variable from Gaussian, binomial, Poisson, etc., distributions, so that it maps onto the linear predictors.

- However, there can be problems when using standard regression approaches applied to datasets with large m (fields) and n (observations).

- Chunking approaches to OLS and generalised regression were illustrated using the biglm and bigglm functions applied to the prescribing data introduced in Chapter 4.

- A starting guess at the coefficients is helpful, and these can be identified by taking a random subset of the data. However, care needs to be taken if using the internal database functions for generating these.

- Some approaches for speeding up GLM parameter estimation for large datasets are suggested.

6.5 QUESTIONING THE ANSWERING PROCESS AND QUESTIONING THE QUESTIONING PROCESS

To conclude this chapter we wish to provide some final advice regarding modelling and testing model assumptions. In particular, we have outlined two key types of inference – classical and Bayesian – and have suggested how these may be used to assess accuracy of model calibration, test a hypothesis or provide probabilistic predictions (although probabilities mean subtly different things for Bayesian and classical inference). The general idea of this chapter has been to provide some guidelines for thinking about models, and how they relate to data. The following chapter will concentrate more on the use of fitting and prediction techniques in a more machine learning framework. Since both approaches appear in practice, it is important to be aware of the ideas underlying each of them.

However, perhaps it is equally important not just to look at the interplay between models and data, but to question the model itself, or the veracity of the data used to calibrate it. In particular, one thing frequently assumed is that the model chosen is actually correct – both Bayesian and classical approaches hinge on this. Some hypothesis tests allow for *nested* alternative models, where one kind of model is a subset of another (e.g. model B may be a subset of regression model A where some of the coefficients are zero). A broader approach may consider several competing models that are not nested – for example, gamma versus Gaussian errors in the example given earlier. There are some techniques that allow this (e.g. the use of Bayes factors), and these may be necessary in some circumstances. A guide is given in Bernardo and Smith (2009).

Similarly, the reliability of the data must also be considered. There are a number of checks (often visual) that may be done to identify outliers. These can be genuine (e.g. a particular house in a database was sold at an incredibly high price) but may also be attributable to recording error (when someone typed in the house price they accidentally hit the trailing zero too many times). In either case it is important to identify them, and scrutinise them to identify, if possible, which kind of outlier they are. Arguably a dataset with a number of genuine outliers needs an appropriate probability distribution to model it, perhaps one with heavier tails than a Gaussian distribution.

Another possibility is that the entire dataset is suspicious. One way of checking this – often used in forensic accounting – is to call on Benford's law (Benford, 1938). This makes use of the observation that in many real-life datasets the first digit in numerical variables is more likely to take a low value. More specifically, if $d_1 \in \{1, \dots, 9\}$ is the first digit, then its probability of occurrence in a large dataset is

$$P(d_1) = \log_{10}\left(1 + \frac{1}{d_1}\right) \tag{6.19}$$

Although not conclusive proof, datasets deviating from this may require further scrutiny. The R package benford.analysis allows these kinds of tests to take place.

In general we advise a sceptical approach, in the sense that nothing can be beyond question. Indeed we would further encourage people to consider the scenarios under which certain questions have arisen – and for which answers are demanded – as well as the consequences of reporting the results of the analysis. Question everything, including the questions and those who ask them.

6.6 SUMMARY

This chapter has aimed to provide an in-depth examination of the process involved in model development. First, it explored models built by utilising ideas from some underlying theory, and then others not built with such definite guidance. It demonstrated that some models can be derived directly via theoretical reasoning about the process being considered, with theory often evolving through empirical observation (as in the van der Waals example), and others can be constructed in a more exploratory way, with the intention of increasing the (possibly subjective) understanding of the process being considered. The key issue in the latter is that the certainty in the form that the model should take is generally less (as in the house price models). This is the situation with much modelling now: the ability to construct models (with less certainty of form) has increased with the many machine learning applications and non-parametric approaches. However, in some cases this has been led to a decrease in consideration of the underlying processes. Second, this chapter highlighted approaches for approximating the degree of certainty/reliability of model results, using coin-tossing examples. These demonstrated how knowing *something* about the inputs allowed probability to be used to indicate the likelihood of different outcomes, both theoretically and through simulation-based examples. This showed how the likelihoods of different answers could be updated if the underlying process changed, the difficulty in determining the probability of precise answers, and the relative ease of being able to calculate probabilities of answers being within a specific range. Third, this chapter illustrated approaches for situations where the precise derivation of probabilities and likelihoods is not possible, using a worked example of competing house price models. Here simulated house price data were constructed from the parameters of simple regression models. The Moran's I (in this case) of the simulated house price data was compared with that of the actual data, thereby generating approximate model probabilities, and allowing one model to be chosen over another. Next, approaches for handling very large datasets in standard linear regression models were illustrated using the `biglm` package and a number of approaches for speeding up GLM parameter estimation for large datasets were suggested. Finally, the importance of taking a critical approach to modelling and the preparedness to question everything was emphasised.

REFERENCES

Anderson, C. (2008) The end of theory: The data deluge makes the scientific method obsolete. *Wired*, 23 June. www.wired.com/science/discoveries/magazine/16-07/pb_theory.

Atkinson, K. A. (1989) *An Introduction to Numerical Analysis.* (2nd edn). New York: John Wiley & Sons.

Benford, F. (1938) The law of anomalous numbers. *Proceedings of the American Philosophical Society*, 78(4), 551–572.

Bernardo, J. M. and Smith, A. F. M. (2009) *Bayesian Theory.* Hoboken, NJ: John Wiley & Sons.

Brunsdon, C. (2015) Quantitative methods I: Reproducible research and human geography. *Progress in Human Geography*, 40(5), 687–696.

Cliff, A. D. and Ord, J. K. (1981) *Spatial Processes: Models & Applications.* London: Pion.

Comber, A. and Wulder, M. (2019) Considering spatiotemporal processes in big data analysis: Insights from remote sensing of land cover and land use. *Transactions in GIS*, 23(5), 879–891.

Freeman, A. M. (1974) On estimating air pollution control benefits from land value studies. *Journal of Environmental Economics and Management*, 1(1), 74–83.

Freeman, A. M. (1979) Hedonic prices, property values and measuring environmental benefits: A survey of the issues. *Scandinavian Journal of Economics*, 81(2), 154–173.

Gale, C. G., Singleton, A., Bates, A. G. and Longley, P. A. (2016) Creating the 2011 Area Classification for Output Areas (2011 OAC). *Journal of Spatial Information Science*, 12, 1–27.

Halvorsen, R. and Pollakowski, H. O. (1981) Choice of functional form for hedonic price equations. *Journal of Urban Economics*, 10(1), 37–49.

Hastie, T. and Tibshirani, R. (1987) Generalized additive models: Some applications. *Journal of the American Statistical Association*, 82(398), 371–386.

Kitchin, R. (2014) Big data and human geography: Opportunities, challenges and risks. *Dialogues in Human Geography*, 3(3), 262–267.

Lancaster, K. J. (1966) A new approach to consumer theory. *Journal of Political Economy*, 74, 132–157.

Lumley, T. (2018) Fast generalised linear models by database sampling and one-step polishing. Preprint, arXiv:1803.05165v1.

McCullagh, P. and Nelder, J. A. (1989) *Generalized Linear Models* (2nd edn). London: Chapman & Hall.

Rosen, S. (1974) Hedonic prices and implicit markets: Product differentiation in pure competition. *Journal of Political Economy*, 82(1), 34–55.

van der Waals, J. D. (1873) On the continuity of the gaseous and liquid states. PhD thesis, Universiteit Leiden.

Vickers, D. and Rees, P. (2007) Creating the UK National Statistics 2001 output area classification. *Journal of the Royal Statistical Society, Series A*, 170(2), 379–403.

Von Bertalanffy, L. (1968) *General System Theory: Foundations, Development, Applications*. New York: George Brazilier.

7

APPLICATIONS OF MACHINE
LEARNING TO SPATIAL DATA

7.1 OVERVIEW

The chapter introduces the application of machine learning to spatial data. It describes and illustrates a number of important considerations, in particular the mechanics of machine learning (data pre-processing, training and validation splits, algorithm tuning) and the distinction between inference and prediction. Some house price and socio-economic data for Liverpool in the UK are introduced and used to illustrate key considerations in the mechanics of machine learning. The classification and regression models are constructed using implementations within the `caret` package, a wrapper for hundreds of machine learning algorithms. Six algorithms are applied to the data to illustrate predictive and inferential modelling:

- Standard linear regression (SLR)/linear discriminant analysis (LDA)
- Bagged regression trees (BRTs)
- Random forests (RF)
- Gradient boosting machines (GBMs)
- Support vector machine (SVM)
- *k*-nearest neighbour (*k*NN).

You will need to load the following packages, some of which may need to be installed with their dependencies:

```
library(sf)
library(tmap)
library(caret)
library(gbm)
library(rpart)
library(tidyverse)
```

```
library(gstat)
library(GGally)
library(visNetwork)
library(rgl)
library(cluster)
library(RColorBrewer)
```

7.2 DATA

Socio-economic data for the Liverpool area are used to illustrate specific consid-erations in machine learning applications. The ch7.RData R binary file includes three sf objects with population census data at two different scales and data of residential properties for sale for the Liverpool area.

The code below loads the file from the internet to your current working directory:

```
# check your current working directory
getwd()
download.file("http://archive.researchdata.leeds.ac.uk/741/1/
              ch7.Rdata",
              "./ch7.RData", mode = "wb")
```

Now load the ch7.RData file to your R/RStudio session and examine the result:

```
load("ch7.RData")
ls()

## [1] "lsoa" "oa" "properties"
```

The three spatial datasets in sf format as briefly introduced in Chapter 6 are as follows:

- oa, a multipolygon object of Output Areas (OAs);

- lsoa, a multipolygon object of Lower Super Output Areas (LSOAs);

- properties, a multipoint object of houses for sale in the Liverpool area.

oa and lsoa represent OAs (~300 people) and LSOAs (~1500 people), respectively. They each contain census data on economic well-being (percentage unemploy-ment, unmplyd), life-stage indicators (percentage of people under 16 years (u16), 16–24 years (u25), 25–44 years (u45), 45–64 years (u65) and over 65 years (o65) and an environmental variable of the percentage of the census area containing green-spaces. The unemployment and age data were from the 2011 UK population census (https://www.nomisweb.co.uk) and the greenspace proportions were extracted from the Ordnance Survey Open Greenspace layer (https://www.ordnancesurvey.co.uk/opendatadownload/products.html). The spatial frameworks were from the EDINA data library (https://borders.ukdataservice.ac.uk). The OAs and LSOAs are both projected in the OSGB projection (EPSG 27700). The OA layer also has a geo-demographic class label (OAC) from the OAC (see Gale et al., 2016) of eight classes as introduced in earlier chapters.

The `properties` data were scraped from Nestoria (https://www.nestoria.co.uk) on 30 May 2019 using their API (https://www.programmableweb.com/api/nestoria). The dataset contains latitude and longitude in decimal degrees, giving location, price in thousands of pounds (£), the number of bedrooms and 38 binary variables indicating the presence of different keywords in the property listing such as 'conservatory', 'garage' and 'wood floor'. The code below lists these:

```
properties %>% st_drop_geometry() %>% select_if(is_logical) %>%
    colnames() %>% paste(" - ",.) %>% paste(collapse='\n') %>% cat()
```

Some EDA can provide an initial understanding of the data, using some of the methods introduced in Chapter 5. For example, the code below creates a ggpairs plot of the oa data, and you could modify this to create a similar plot of the lsoa data. Here again you can widen the plot display pane in RStudio if the ggpairs plot does not display:

```
# OA pairs plot
oa %>% st_drop_geometry() %>% select_if(is.numeric) %>%
    ggpairs(lower = list(continuous =
                          wrap("points", alpha = 0.5, size=0.1))) +
    theme(axis.line=element_blank(),
          axis.text=element_blank(),
          axis.ticks=element_blank()))
```

If you examine the oa and lsoa correlations you should note a number of things:

1. the broad similarity of the distribution of attribute values across the two scales;

2. the extremely skewed distributions of the greenspace attribute (gs_area) at both scales, but with less skew at the coarser LSOA scale;

3. the relatively large correlations between employment (employd) and age group percentages;

4. stronger correlations at OA level.

These suggest that the modifiable areal unit problem (Openshaw, 1984) might cause some differences in model outcome in this case.

Alternative visualisations are possible with the `properties` data:

```
ggplot(properties,aes(x=Beds,y=Price, group=Beds)) + geom_boxplot()
```

We can also use `tmap` to see the geographical distribution of houses with different numbers of bedrooms, which has a geographical pattern:

```
tmap_mode('view')
tm_shape(properties) +
   tm_dots(col='Beds', size = 0.03, palette = "PuBu")
tmap_mode('plot')
```

However, given the complexity of the variables and their binary nature (try running `summary(properties)`), perhaps a more informative approach to understanding structure in the properties data is to use a regression tree. These are like decision tress with a hierarchical structure. They use a series of binary rules to try to partition data in order to determine or predict an outcome. If you have played the games *20 Questions* or *Guess Who* (https://en.wikipedia.org/wiki/Guess_Who%3F) then you will have constructed your own decision tree to get to the answer. In *Guess Who*, an optimal strategy is to try to split the potential solutions by asking questions that best split the remain potential answers, and regression trees try to do the same thing. They have a *root node* at the top which performs the first split, and branches connect to other nodes, with all branches relating to a yes or no answer connecting to any subsequent intermediary nodes (and associated division of potential outcomes) before connecting to *terminal nodes* with predicted outcomes. Regression trees use *recursive partitioning* to generate multiple subsets such that the members of each of the final subsets are as homogeneous as possible, and the results describe a series of rules that were used to create the subsets/partitions. This process can be undertaken for both continuous variables using *regression* trees and categorical variables using *classification* trees. There are many methodologies for constructing these, but one of the oldest is known as the *classification and regression tree* approach developed by Breiman (1984).

A regression tree for house price can be constructed using the `rpart` package. Note the use of `st_drop_geometry` and `mutate_if` to convert the logical TRUE and FALSE keyword values to character format:

```
set.seed(78910) # for reproducibility
tree_model <- rpart(Price~.,data=properties %>%
                    st_drop_geometry() %>%
                    mutate_if(is_logical,as.character))
```

You can examine the regression tree, noting that indented information under each split represents a subtree:

```
tree_model
```

This can also be visualised using the visNetwork package as in Figure 7.1, which shows the variables that are used to split the data, the split values for the variables, and the number of records that are partitioned by the branch thickness. The main branches in determining price are related to the number of bedrooms, whether en suite facilities are available, and whether it is Victorian. Interestingly, it is evident that some geography comes into play for larger semi-detached houses. However, the tree gives an indication of the data structure in relation to the Price variable:

```
library(visNetwork)
visTree(tree_model,legend=FALSE,collapse=TRUE,direction='LR')

visTree(tree_model,legend=FALSE,collapse=TRUE,direction='LR')
```

Key Points

- Socio-economic data at two different scales plus data on residential properties for sale for the Liverpool area were loaded into the R session.

- Some EDA and spatial EDA were used to examine the data, using methods from Chapter 5.

- These were enhanced with a regression tree approach to examine the structure of the properties data relative to house price.

7.3 PREDICTION VERSUS INFERENCE

Statistical models are constructed for two primary purposes: to predict or to infer some process. The distinction between the two activities has sometimes been blurred as the volume of activities in data analytics (or statistics – see the preface to this book) has increased, and the background training of those involved in these activities has diversified. Sometimes the terminology is confusing, especially the use of the term *inference* or *infer*: in the previous chapter we discussed how one could *infer* from classical and Bayesian models, for example.

In truth, both prediction and inference seek to *infer* y from x in some way through some function f, where x represents the input (predictor, independent, etc.) variables, y is the target (or dependent) variable and ϵ is some random error term:

$$y = f(x) + \epsilon \tag{7.1}$$

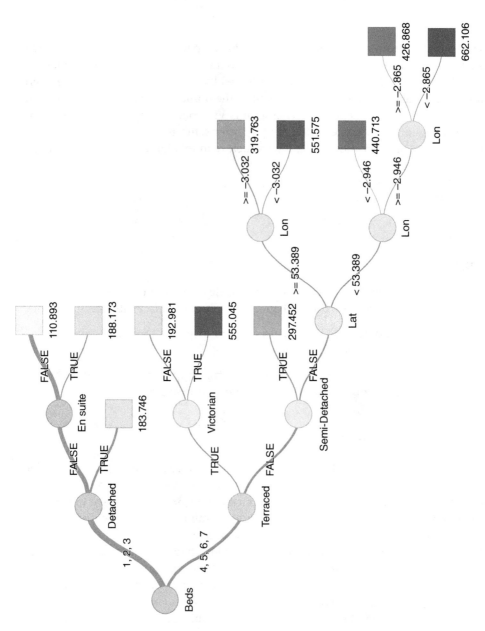

Figure 7.1 The regression tree of house price in the properties data

It is always useful to return to core knowledge to (re-)establish fundamental paradigms, and in this case to consider how others regard the distinction between prediction and other kinds of inference:

> in a real estate setting, one may seek to relate values of homes to inputs such as crime rate, zoning, distance from a river, air quality, schools, income level of community, size of houses, and so forth. In this case one might be interested in how the individual input variables affect the prices – that is, *how much extra will a house be worth if it has a view of the river?* This is an **inference problem**. Alternatively, one may simply be interested in predicting the value of a home given its characteristics: *is this house under- or over-valued?* This is a **prediction problem**. (James et al., 2013, p. 20; bold emphasis added)

Viewed through this lens, prediction and inference can be distinguished as follows.

Prediction uses the estimated function f to forecast y, for example over unsampled areas when data for those areas become available, or into the future, given different future states.

Inference uses the estimated function f to understand the impact of the inputs x on the outcome y. It is often associated with process understanding, with the aim of, for example, determining how much of y is explained by a particular x, for instance x_i, through its partial derivative $\partial f/\partial x_i$.

The aim in both cases is to identify the function f that best approximates the relationships or structure in the data. And, although prediction and inference are both in a sense *inferential*, with one focusing on forecasting and the other on process understanding, and both may use the same estimation procedure to determine f, the major difference is that they have different procedural considerations. This is especially the case for how the issues around independence and collinearity among predictor variables are treated, for example: highly correlated input variables may make it difficult to separate variable effects when the objective is inference, whereas this does not matter for prediction, which is concerned only with the accuracy and reliability of the forecast.

Many of the examples in this chapter can be thought of as examples of *machine learning*. Typically, the main objective in these is *reliable* prediction. That is, the underlying need is to predict some quantity. In order to create reliable predictions it can be argued that some understanding of process may be helpful – but in situations where that proves difficult, or in urgent situations where predictions are required quickly, there are some techniques that 'short-circuit' the understanding of causality, and attempt to predict solely on patterns in the data. A mixture of approaches will be considered here.

Key Points

- Statistical models are constructed for prediction or for inference (understanding).

- It is important to distinguish between these two activities as the terminology can be confusing.

- Prediction creates a model and uses this to forecast or predict using unseen data.

- Inference creates a model in order to understand the relationship between some process and different explanatory factors.

- Prediction and inference have different procedural considerations, for example around issues of variable independence and collinearity.

- In machine learning the objective is typically prediction rather than inference.

7.4 THE MECHANICS OF MACHINE LEARNING

Recall that the basic objectives of machine learning are process understanding (inference) and prediction. Machine learning achieves these by determining data structure (relationships) and identifying the functions that best approximate these relationships.

There are two general approaches to machine learning: supervised and unsupervised learning.

Supervised learning is undertaken when the outcomes are known, and prior knowledge exists of what the model or function output values should be. More formally, for each observation of the predictor variables x_i the associated response y_i is known. In this situation, the aim is to determine the function f that best approximates the relationship between the *observed* outputs and inputs, by predicting the *known* response variable as accurately as possible in order to best predict future or unknown responses. The overall approach is to split the data into training and validation data subsets, to construct the model using the training data, to then apply (test) the model over the validation data for which the actual observed outcomes are known, and finally to compare the observed with predicted values to determine the model reliability or fit (see below).

Unsupervised learning describes the situation when the response is unknown. That is, for each observation, x_i is observed but not the response y_i. It is *unsupervised* because the there is no response to supervise the fitting of a function. This limits the statistical analyses that are possible to those that seek to understand the relationships between the variables over observations, such as cluster analysis or classification. Here all the data are used as inputs to construct the model, and measures of model fit describe the in-sample prediction accuracy. In classification models, model fit is determined by evaluating the within-class homogeneity – that is, a summary of the closeness (similarity) of each observation (record) to the class to which it is allocated in the clustering process.

In both supervised and unsupervised approaches the goal is to identify specific relationships or structure in the predictor variables data in order to generate a robust and powerful statistical model. There are additional data considerations such as whether or not to rescale the data, which variables to use as inputs (although the latter is more important for inference in supervised approaches), model specification (variable selection) in supervised approaches, determining the number of clusters k in unsupervised classification approaches, model evaluation, and so on. Much of the actual development and application of machine learning is concerned with these considerations rather than actually running the algorithm. These are described and illustrated in the subsections below.

7.4.1 Data rescaling and normalisation

Many statistical machine learning models use some form of multivariate distance measure to determine how *far apart* different observations are from each other. A multivariate feature space is defined by the predictor variables passed to the model. Consider linear regression. It determines the hyperplane with the minimum summed distance between each observation and the plane. By way of example, a very simple linear regression model of Price in the properties data can be constructed from Beds and Lat. Because we have only two predictor variables the model hyperplane and its relationship with the target variable (Price) can be visualised using a 3D plot. The code below should generate a pop-out window that you can manipulate (rotate, zoom, etc.):

```
library(rgl)
df <- properties %>% st_drop_geometry() %>%
    select(Price, Beds, Lat) %>%
  mutate(Beds = as.numeric(as.character(Beds)))
plot3d(lm(Price~Beds+Lat, data = df),col="orange",type = 's',
        size=1,vars = df[,c("Price", "Beds", "Lat")])
# run the code below to see the plot in your RStudio session
# rglwidget(width = 600, height = 1100, reuse = FALSE)
```

However, the problem in this case (and in most others) is that the variables are in different units of measurement: Lat is in degrees, Beds is a count of beds and Price is in thousands of pounds. Floor area, if present, would be in square metres. The danger is that models constructed on the *raw* data can be dominated by variables with large numerical values. For example, if Price were in pounds rather than thousands of pounds then this domination would be greatly enhanced. Similarly, the relative influence of Lat would change if it were expressed as an easting in metres, as in the OSGB projections of oa and lsoa. However, this seems to be a somewhat arbitrary way to proceed – the houses have the same physical size, and the same location regardless of units of measurement, and it seems non-sensical for the notion of *closeness* to change with any change in these units.

For this reason, data are often rescaled (or scaled) prior to applying maximum likelihood algorithms. Usually this is done either by computing z-scores, or rescaling by minimum and maximum values. For any variable x, rescaling by z-score is as follows:

$$z = \frac{x - \bar{x}}{\sigma_x} \tag{7.2}$$

where $\bar{x}$ is the mean of x and σ_x is the standard deviation of x. The result is that z has a mean of 0 and a standard deviation of 1, regardless of the units of x. Additionally, the distributions of the z-scores are the same as those of x.

An alternative approach is scaling via minimum and maximum values to generate a rescaled variable u:

$$u = \frac{x - x_{min}}{x_{max} - x_{min}} \tag{7.3}$$

This maps the largest value of x to 1, and the smallest to 0. Again the distributions are unchanged and the values are independent of the measurement units.

R has a number of in-built functions for rescaling data, including the `scale` function in the base R installation which rescales via z-scores. This changes the data values but not their distribution:

```
library(dplyr)
df <- properties %>% st_drop_geometry() %>%
    select(Price, Beds, Lat) %>%
   mutate(Beds = as.numeric(as.character(Beds)))

X <- df %>% scale()
```

The object X now contains z-score-rescaled columns from df – you can check the first six observations and compare with the original data:

```
head(df)
head(X)
```

You can also check that this has been applied individually to each variable and not in a 'group-wise' way across all of the numeric variables using a single pooled mean and standard deviation across all variables:

```
apply(X, 2, mean)
apply(X, 2, sd)
```

The caret package also includes parameters for rescaling (transforming) the input data using the means and standard deviations as part of the call to the functions that create the models, as will be demonstrated later. It also contains the preProcess() function which supports transformation at model creation (training) time. It does not actually transform the data, rather it suggests transformations:

```
library(caret)
props_pp =
    properties %>% st_drop_geometry() %>%
    mutate_if(is_logical, as.character) %>%
    preProcess(method =
                c("center", "scale", "YeoJohnson", "nzv", "pca"))
```

The outputs can be examined in order to determine which variables were ignored, centred, or scaled, for example, whether to undertake a PCA or not (which looks very promising in this case), and so on:

```
props_pp
props_pp$method
```

7.4.2 Training data

Supervised machine learning for prediction requires the input data to be split into two subsets: one to create the model and the other test or evaluate its performance.

In supervised machine learning, the data contain the target variable (output) and this allows the model to be evaluated using a quantifiable and objective metric. Recall that the aim is to use a matrix of predictor variables $x_{i...n}$, now denoted by X, and a response variable to be predicted y in order to determine some function $f(X)$ such that $f(X)$ is as close to y as possible.

Thus, one of the critical issues in supervised machine learning and the associated training and validation splits is that any given split of the data may not be representative of the dataset as a whole (and therefore our knowledge of the problem) and models constructed on such a subset may lead to erroneous prediction.

The caret package includes the createDataPartition() function that splits data into two sets for training and validation and ensures that the distribution of outcome variable classes is similar in both. This is because it uses a bootstrapping approach (random sampling with replacement) to create a stratified random split. A random seed is set below to ensure reproducibility:

```
set.seed(1234)
train.index = as.vector(createDataPartition(
    properties$Price, p = 0.7, list = F))
```

The distributions of the target variable `Price` are similar across the two splits:

```
summary(properties$Price[train.index])
summary(properties$Price[-train.index])
```

The split or partition index can be used to create the training and test data as in the code snippets below:

```
train_data =
  properties[train.index,] %>% st_drop_geometry() %>%
  mutate_if(is_logical,as.character)
test_data =
  properties[-train.index,] %>% st_drop_geometry() %>%
  mutate_if(is_logical,as.character)
```

7.4.3 Measures of fit

The identification of the function f in equation (7.1) depends not only on the training data, but also on a number of tuning parameters. Varying the tuning parameters will alter how close $f(X)$ is to y, with the aim of identifying which combination of parameters get the closest. A key tuning parameter is the choice over measures of fit.

The general method of evaluating model performance using this split is called cross-validation (CV). This is an out-of-sample test to validate the model using some fitness measure to determine how well the model will generalise (i.e. predict using an independent dataset).

So the general method of evaluating model performance is to split the data into a training set and a test set, denoted by $S = X$, y and $S' = X'$, y' respectively, and calibrate f using S, with a given set of tuning parameters. Then X' is used to compute $f(X')$ and the results are compared to y'. This CV procedure returns a measure of how close the predictions of y' are to the observed values of y' when $f(X')$ is applied.

There are a number of ways of measuring this. Two commonly used methods (for regression) are the root mean square error (RMSE), defined by squaring the errors (the residual differences between y' and $f(X')$), finding their mean and taking the square root of the resulting mean:

$$\text{RMSE} = \sqrt{\sum \frac{(y' - f(X'))^2}{n}} \tag{7.4}$$

An alternative is the mean absolute error (MAE), defined by

$$\text{MAE} = \sum \frac{|y' - f(X')|}{n} \tag{7.5}$$

And perhaps the most common accuracy measure for regression model fit is R^2, the coefficient of determination, calculated from the residual sum of squares over the total sum of squares:

$$R^2 = 1 - \frac{\Sigma(y' - f(X'))^2}{\Sigma(y' - \bar{y}')^2} \qquad (7.6)$$

All of these essentially measure the degree to which the predicted responses differ from the actual ones in the test dataset. For the MAE and RMSE measures, smaller values imply better prediction performance, and for R^2 higher ones do. Similarly, in categorical prediction, the proportion of correctly predicted categories is a useful score, and again larger values imply better performance.

Of course the descriptions above relate to models applied to test (validation) data subsets, typically in the context of prediction. Models are also evaluated during their construction using *leave-one-out cross-validation* and *k*-fold cross-validation. These procedures include a resampling process that splits the data into *k* groups or 'folds'. The approach, in outline, is to randomly split the data into *k* folds, such that each observation is included once in each fold. Then, for each fold, keep one as the hold-out or test data, train the model on the rest and evaluate it on the test, retaining the evaluation score. The retained evaluation scores are summarised to give an overall measure of fit. In this process each individual observation is uniquely assigned to a single fold, used in the hold-out set once and used to train the model $k - 1$ times. It takes longer than fitting a single model as it effectively undertakes $k + 1$ model fits, but is important for generating reliable models, whether for prediction or inference.

The caret package includes a large number of options for evaluating model fit that can be specified using the trainControl() function. You should examine the method parameters that can be passed to trainControl in the help. The trainControl() function essentially specifies the type of 'in-model' sampling and evaluation undertaken to iteratively refine the model. It generates a list of parameters that are passed to the train function that creates the model:

```
ctrl1 <- trainControl(method = "cv",
                      number = 10)  # a 10-fold CV
```

Others are available, such as repeated *k*-fold cross-validation and leave-one-out. The trainControl function can be used to specify the type of resampling:

```
# a 10-fold repeated CV, repeated 10 times
ctrl2 <- trainControl(method = "repeatedcv",
                      number = 10,
                      repeats = 10)
```

The outputs of these could be examined one by one but in this case only the first three values are different:

```
# there are many settings
names(ctrl1)
# compare the ones that have been specified
c(ctrl1[1], ctrl1[2], ctrl1[3])
c(ctrl2[1], ctrl2[2], ctrl2[3])
```

7.4.4 Model tuning

So far, we have considered a number of aspects associated with the mechanics of machine learning (data preparation, training data, measures of fit), and in each case we have noted that there is a caret function to do this, or a parameter that can be specified and passed to a function that creates or trains the model. The train function in caret has not been introduced yet but we encounter it below in the context of model refinement or *tuning*.

Different machine learning approaches require different *tuning parameters*. Some relate to the input data as described above, other *hyperparameters* relate to the model configuration and control the model training process. For example, in a neural network these include the number of layers, the number of nodes in each layer, as well as the fitness measures and other parameters specified in trainControl.

The parameters for any specific model can be *optimised* (tuned) by adjustment until a best fit is found through the accuracy of the trained model. In all cases the objective of tuning is to modify the model and find the best combination of parameters for the task in hand. The supporting documentation for caret provides an example of tuning (see https://topepo.github.io/caret/model-training-and-tuning.html). This describes the steps in tuning as:

```
 1. Define sets of model paramaters to optimise (tune)
 2. for each parameter set do
 3. | for each resampling iteration do
 4. | | hold out specific samples
 5. | | fit the model on the remainder
 6. | | predict on the hold-out samples
 7. | end
 8. | Calculate the fitness from hold-out samples
 9. end
10. Determine the set of optimal parameters
11. Fit final model to all training data with optimal parameters
```

The specific tuning parameters for any given caret machine learning model can be listed using the modelLookup function. The code below describes these for a *k*-nearest neighbour (*k*NN) model:

```
modelLookup("knn")

##    model parameter       label forReg forClass probModel
## 1 knn              k  #Neighbors   TRUE     TRUE      TRUE
```

This indicates that only k, the number of neighbours, needs to be optimised. It also indicates that the model can be used for regression and classification, and that the model produces class probabilities. We will use this model as a simple example of the tuning process and later on a complex example will be illustrated using a gradient boosting machine approach.

Model choice should be determined by the objective of the study. In this case, the objective is to construct a predictive model of house price and there is a reasonable argument for using a nearest neighbour approach. It could be argued that when suggesting the 'asking price' for houses being put on sale, many estate agents would look at the prices of similar-sized houses nearby. This is more or less mirrored by the nearest neighbour algorithm.

Before going into the detail, it is instructive to fit an initial model:

```
set.seed(123) # for reproducibility
knnFit <- train(Price ~ ., data = train_data, method = "knn")

knnFit <- train(Price ~ ., data = train_data, method = "knn")
```

We can inspect the outputs, which tell us that the model optimised on nine nearest neighbours: remember that these are neighbours in a multidimensional feature space, which also includes geographic space as two of the inputs are latitude and longitude. These are shown in the error from the resampling for RMSE, MAE and R^2:

```
knnFit

## k-Nearest Neighbors
##
## 2963 samples
##   41 predictor
##
## No pre-processing
## Resampling: Bootstrapped (25 reps)
## Summary of sample sizes: 2963, 2963, 2963, 2963, 2963, 2963, ...
## Resampling results across tuning parameters:
##
## k RMSE       Rsquared   MAE
## 5 121.4402 0.4558293 65.39469
## 7 117.4175 0.4798227 63.10847
## 9 116.0912 0.4874645 62.31709
##
```

```
## RMSE was used to select the optimal model using
##   the smallest value.
## The final value used for the model was k = 9.
```

Note also that much more detail is included in the result: try entering names(knnFit) and examine the help for the caret train function for descriptions of these:

```
names(knnFit)
help(train, package = "caret")
```

The fits can be plotted:

```
ggplot(knnFit)
ggplot(knnFit,
metric = "MAE")
```

It is clear that the error rates have not bottomed out and that we might be able to improve the model fit by letting it tune for a bit longer. This can be done by passing additional parameters to the train() function such as tuneLength. What this does is make the algorithm evaluate a greater number of parameter settings:

```
set.seed(123)
knnFit <- train(Price ~ ., data = train_data, method = "knn",
                tuneLength = 20)
```

In this case a greater number of values of k are explored. Examine the outputs as before, and notice that the value of k is now 27, indicating the differences between the measures of model fit given by the different metrics. The train argument metric defaults to RMSE, but others can be specified. The different fits can be visualised, but also note that whereas RMSE saturated at around 25, the curves for MAE and R^2 potentially suggest a different value (Figure 7.2).

```
knnFit$results %>% select(k, RMSE, MAE, Rsquared) %>%
  pivot_longer(-k) %>%
  ggplot(aes(x = k, y = value)) +
  geom_point() +
  geom_line()+ ylab("Accuracy") +
  facet_wrap("name", scales = "free")
```

To examine this impact of the metric argument, the code below specifies MAE:

```
set.seed(123)
knnFit <- train(Price ~ ., data = train_data, method = "knn",
                tuneLength = 20, metric = "MAE")
```

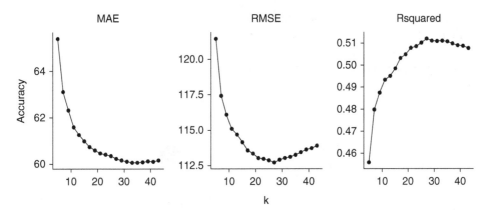

Figure 7.2   Model fits under different metrics with different numbers of neighbours

Note that the best-fitting model is for a value of *k* of around 33:

```
ggplot(knnFit, metric = "MAE")
```

This is *reproducible* because of the use of the set.seed() function. The potential for ambiguity and variation in the results of the train function arises because of the random element of the processing: where in the data the algorithm starts, when fitting the data. The set.seed function ensures repeatable results. Try running the code snippet defining knnFit above without set.seed(123) and examine the optimal values of *k* that are determined in different runs.

Finally, robust training and validation splits *within the training* are needed to ensure that the model is not a function of the split but of the general patterns in the data. As well as splits being done outside of the train() function to set up training and validation subsets, it is also possible *within* the function as discussed above, through the trainControl() function which specifies training parameters and stabilises the results. As a rule of thumb 10 folds with 10 repeats should result in a robust model. This takes a bit longer to run, but the outputs should be more stable, avoiding the need to (arbitrarily) pass a value to set.seed in order to generate reproducible results:

```
ctrl <- trainControl(method = "repeatedcv",
                     number = 10, repeats = 10)
knnFit <- train(Price ~ ., data = train_data, method = "knn",
                tuneLength = 20, metric = "MAE", trControl = ctrl)
```

If the model is inspected again then the model properties can be compared with earlier models:

```
knnFit
```

A further important consideration is ability to normalise the data within the function (rather than using the scale function to standardise the data with z-scores before input). This can be done through the preProcess parameter allowing the original data to be used as input to training and later on using the predict function:

```
knnFit <- train(Price ~ ., data = train_data, method = "knn",
                tuneLength = 20, metric = "MAE", trControl = ctrl,
                preProcess = c("center","scale"))
knnFit
```

The *k*NN model has only one parameter to tune. Typically many more need to be evaluated. This is illustrated with a gradient boosting machine (GBM) model which is described later in this chapter. The tunable parameters for the GBM can again be listed with the modelLookup function. Note that caret does not load all the functions and packages on installation, but instead assumes that they are present and loads them as needed: if a package is not present, caret provides a prompt to install it.

```
modelLookup("gbm")

##    model             parameter                    label  forReg
## 1    gbm               n.trees    # Boosting Iterations    TRUE
## 2    gbm   interaction.depth          Max Tree Depth        TRUE
## 3    gbm             shrinkage             Shrinkage        TRUE
## 4    gbm       n.minobsinnode   Min.  Terminal Node Size    TRUE
##     forClass  probModel
## 1       TRUE       TRUE
## 2       TRUE       TRUE
## 3       TRUE       TRUE
## 4       TRUE       TRUE
```

This tells us that there are four tuning parameters for a GBM:

- n_trees, the number of trees;
- interaction.depth, a tree complexity parameter;
- shrinkage, the learning/adaptation rate;
- n.minobsinnode, the minimum number of samples in a node for splitting.

Again an initial model can be generated and the results evaluated, noting that you may have to install the gbm package and that the model may take some time to train:

```
set.seed(123)
ctrl <- trainControl(method="repeatedcv", repeats=10)
gbmFit <- train(Price ~ ., data = train_data, method = "gbm",
                trControl = ctrl, verbose = FALSE)
gbmFit
```

The results show that the different accuracy measures (RMSE, R^2 and MAE) provide an indication of the different agreements averaged over the cross-validation iterations in ctrl. However, they also indicate that the default GBM implementation in caret evaluates different values for only two tuning parameters (interaction.depth and n.tree) and uses only a single value for shrinkage (0.1) and n.minobsinnode (10). In this case the best model was found with n.trees = 50, interaction.depth = 3, shrinkage = 0.1 and n.minobsinnode = 10.

Here it is more evident than in the *k*NN case that the train function automatically evaluates a limited set of tuning parameters and since, by default, the size of this is $3^{Parameters}$, this results in 9 being evaluated in the case of GBM (compared with 3 for *k*NN).

However, as well as increasing the tuneLength passed to train as above, it is possible for user-defined sets of parameters be evaluated. These are passed to the tuneGrid argument in train, for which a series of models are created and then evaluated. This requires a data frame to be created and passed to the tuneGrid argument in train, with columns for each tuning parameter, named in the same way as the arguments required by the function. The examples in the information box below define a grid that contains 270 combinations of parameters – these will take an hour or two to evaluate.

Returning to the gbmFit object defined above, the results can be examined by printing out the whole model or just the best tuning combinations of those evaluated under the default settings:

```
gbmFit
gbmFit$bestTune
```

The best model in this case has n.trees = 150, interaction.depth = 3, shrinkage = 0.1 and n.minobsinnode = 10. This can be accessed directly using the which.min and which.max functions as below to interrogate the different accuracy measures. In this case they all suggest the same optimal parameters:

```
gbmFit$results[which.min(gbmFit$results$RMSE),]
gbmFit$results[which.max(gbmFit$results$Rsquared),]
gbmFit$results[which.min(gbmFit$results$MAE),]
```

Expanding the tune grid

The following code snippet defines a grid that contains 270 combinations of parameters – these will take an hour or two to evaluate:

```
params <- expand.grid(n.trees = seq(50, 300, by = 50),
                      interaction.depth = seq(1, 5, by = 1),
                      shrinkage = seq(0.1, 0.3, by = 0.1),
                      n.minobsinnode = seq(5,15,5))
dim(params)
head(params)
gbmFit <- train(Price ~ ., data = train_data, method = "gbm",
                trControl = ctrl,
                tuneGrid = params, verbose = FALSE)
```

The results can be examined:

```
gbmFit$bestTune
```

Alternatively, the whole model can be printed:

```
gbmFit
```

The best model has n.trees = 300, interaction.depth = 5, shrinkage = 0.2 and n.minobsinnode = 15. This can be accessed directly using the which.min and which.max functions as below to interrogate the different accuracy measures. Notice that the MAE provides slightly different optimal parameters:

```
gbmFit$results[which.min(gbmFit$results$RMSE),]
gbmFit$results[which.max(gbmFit$results$Rsquared),]
gbmFit$results[which.min(gbmFit$results$MAE),]
```

If you are unsure whether the full range of tuning parameters has been evaluated you could define a grid containing even more combinations of parameters, perhaps leaving this to run overnight:

```
params <- expand.grid(n.trees = seq(50, 400, by = 50),
                      interaction.depth = seq(1, 5, by = 1),
                      shrinkage = seq(0.1, 0.5, by = 0.1),
                      n.minobsinnode = seq(10,50,10))
dim(params)
#head(params)
gbmFit <- train(Price ~ ., data = train_data, method = "gbm",
                trControl = ctrl,
                tuneGrid = params, verbose = FALSE)
```

7.4.5 Validation

Finally, having created trained a well-tuned model, it can be applied to the hold-out data for model evaluation using the `predict` function. This gives an indication of the *generalisability* of the model – the degree to which it is specific to the training/validation split. Theoretically the split was optimised such that the properties of the response variable (`Price`) were the same in both subsets. The code below goes back to the last *k*NN model, applies this to the test data and compares the predicted house price values with the actual observed house price values. The scatter plot of these with some trend lines is shown in Figure 7.3.

```
pred = predict(knnFit,newdata = test_data)
data.frame(Predicted = pred, Observed = test_data$Price) %>%
   ggplot(aes(x = Observed, y = Predicted))+
   geom_point(size = 1, alpha = 0.5)+
   geom_smooth(method = "loess", col = "red")+
   geom_smooth(method = "lm")
```

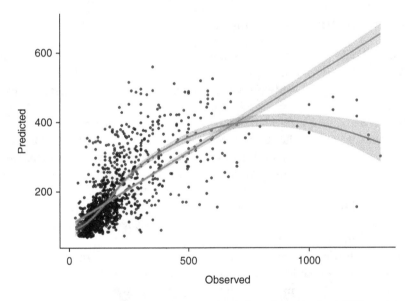

Figure 7.3   Predicted against observed house price values from the final *k*NN model, with the model fit indicated by the blue linear trend line and the loess trend in red giving an indication of the variation in prediction from this

Both loess and linear regression trend line are included in Figure 7.3. We have not fitted a non-linear regression, but this trend line reflects the variation in the predicted and observed data overall. The `loess` inflections suggest that the model does reasonably well at predicting house prices in the lower middle of the range of values, where both the loess and lm (linear model) trend lines are parallel, but it does less well at predicting higher observed values.

A measure of overall model fit can be obtained, which essentially summarises the comparison of the observed values of Price against the predicted ones:

```
postResample(pred = pred, obs = test_data$Price)

##          RMSE     Rsquared          MAE
## 115.8387874    0.4636135   65.3384096
```

7.4.6 Summary of key points

The basic objectives of machine learning are process understanding (inference) and prediction. These are pursued using two general approaches:

- Supervised learning, which is typically used for prediction on the full dataset and using data split into training and testing (validation) subsets.

- Unsupervised learning, which is typically used for inference (understanding) or classification and uses the full dataset with model fit measures describing in-sample accuracy.

Input data should be rescaled (normalised) in order to minimise the scale differences between different units of measurement in fitting the model (variables with large values can unreasonably dominate model fitting). There are different ways to do this but the standard approach is to use z-scores that transform the data to have a mean of 0 and a standard deviation of 1, regardless of the units of x.

For supervised approaches to prediction, input data need to be split into two subsets (one to create the model, the other to evaluate its performance) and there are functions for ensuring that the target variable has a similar distribution in both subsets.

A number of different measures of fit can be used to evaluate model performance, including R^2, root mean square error (RMSE) and mean absolute error (MAE). They all measure the degree to which the predicted values differ from actual ones in the test dataset.

Model fitting uses different in-sample (i.e. training data) cross-validation folding procedures in which individual data observations are repeatedly held out and predicted to generate the model. In the caret package the trainControl function specifies the nature of the 'in-model' sampling and evaluation.

Models should be tuned by exploring combinations of model parameters. The model implementations in the caret package automatically examine a small subset of tuning parameters. However, a larger tuning grid can be specified, by either increasing tuneLength or specifying a tuning grid to be passed to tuneGrid in train.

The aim of validation is to determine the *generalisability* of a predictive model – how well it predicts. It is undertaken by applying the model to data whose results are known and comparing the predicted value with the observed value.

7.5 MACHINE LEARNING IN caret

The previous section on the *k*NN and GBM models highlighted the generic considerations when undertaking machine learning: data transformations, training data, measures of fit, tuning parameters and validation.

This section illustrates a small number of different machine learning algorithms. There are a number of R packages that support machine learning, perhaps the best developed of which is the caret package (*classification and regression training*; Kuhn, 2015), as used in the illustrations above. The advantage of caret is that it provides a consistent environment for an integrated interface to many machine learning algorithms (see https://topepo.github.io/caret/available-models.html or enter names(getModelInfo()) in the R console to see the full list). The different algorithms are contained in other R packages, and caret provides a bridge to them and standardises a number of common tasks. It contains tools for data splitting and pre-processing, model tuning, determining estimates of variable importance and related functionality. Details of the caret package (Kuhn, 2015) can be found at http://topepo.github.io/caret/ and in the book by the package author (Kuhn and Johnson, 2013). Recall that caret does not load all packages it links to on installation, but will prompt you to install any package you use with caret if it is not present.

A number of different algorithms (approaches) are used for prediction, inference and classification:

- Standard linear regression/linear discriminant analysis as benchmark
- *k*-nearest neighbour
- Bagged regression trees
- Random forests
- Gradient boosting machines
- Support vector machine.

These are all applied to the same data, the creation of which is described in the next subsection.

7.5.1 Data

The code below links the properties data to the socio-economic data in the Output Area layer (oa) by location. This is a spatial overlay/intersection operation – see Brunsdon and Comber (2018) for a full description of the various spatial integration operations that are possible in R. The result has the census data attributes attached to each record in properties based on the record's location. An alternative would be to pool the house price data over the census areas, but this would require the house price data to be averaged in some way (e.g. average house price or the percentage of houses with *n* bedrooms), and would lose much of the nuance in the data such as the keywords indicating whether the property had a garage, a conservatory or was semi-detached.

To spatially intersect spatial data layers they need to be in the same geographic projection. The code below converts the properties data to the OSGB projection, with Easting and Northing in metres replacing the Lon and Lat variables in degrees. The two datasets are then spatially intersected and a few unwanted variables are removed (census area codes, latitude, longitude):

```
properties %>% st_transform(27700) %>%
  # add Easting and Northing as variables
  mutate(Easting = (properties %>% st_coordinates() %>% .[,1]))  %>%
  mutate(Northing= (properties %>% st_coordinates() %>% .[,2])) %>%
  # intersect with OA data and drop unwanted variables
  st_intersection(oa) %>%
  # drop Lat, Lon, OAC code, LSOA ID and sf geometry
  select(-Lon, -Lat, -OAC, -code) %>%
  st_drop_geometry() %>%
  # remove NAs and pipe to a data.frame
  drop_na() -> data_anal
```

The structure of the new dataset can be examined using regression trees as before:

```
set.seed(123)
tree_model_oa <- rpart(Price~.,data=data_anal %>%
                       mutate_if(is_logical,as.character))
```

This can be visualised as before using visTree() in the visNetwork package:

```
visTree(tree_model_oa,legend=F,collapse=TRUE,direction='LR')
```

When the results are examined, the impact of the socio-economic variables is clear at the third level of branches. The Beds variable is still the driving factor on price, with unmplyd, Northing and Detached down one main branch, and Northing and u45 down the other. The properties keywords Detached and Kitchen are towards the end of the branches. This highlights the importance of local geographical context in explaining and therefore understanding structure (relationships) in data. Northing especially is present in many branches.

Also note that the data now cover just the northern side of the River Mersey, dropping the data on the Wirrel (to the south of the river), as the census data cover the metropolitan area of Liverpool. The data have now been reduced by nearly 50% from 4230 to 2211 observations.

The new dataset needs to be split into training and validation (testing) subsets:

```
set.seed(1234)
train.index =
    as.vector(createDataPartition(data_anal$Price, p = 0.7, list = F))
train_anal = data_anal[train.index,]
test_anal = data_anal[-train.index,]
```

The training and validation subsets can be scaled, remembering to keep the target variable in its original form:

```
train_z =
    train_anal %>% select(-Price) %>%
    mutate_if(is_logical,as.character) %>%
    mutate_if(is_double,scale) %>% data.frame()
test_z =
    test_anal %>% select(-Price) %>%
    mutate_if(is_logical,as.character) %>%
    mutate_if(is_double,scale) %>% data.frame()
# add unscaled price back
train_z$Price = train_anal$Price
test_z$Price = test_anal$Price
```

Note that it is important to rescale the data with z-scores, *after* splitting the data into training and validation subsets. The danger of normalising/rescaling the data *before* the split is that information about the *future* is being introduced into the training explanatory variables.

These data will be used in the `caret` machine learning prediction models below. Inference models will use the full data and the classification models will explore the OAC variable, and so new data will be created for these.

7.5.2 Model overviews

A sample of the many possible machine learning algorithms and models for classification and regression are illustrated. These have been selected to represent a spectrum of increasing non-linear complexity through ensemble approaches such as bagged regression trees, random forests and learning approaches (support vector machines), as well as classic approaches such as k-nearest neighbour and discriminant analysis. This subsection provides a (very) brief overview of these. In subsequent sections and subsections their strengths of prediction, inference and classification are compared with standard linear regression/linear discriminant analysis. Recall that `caret` will evaluate a small range of tuning parameters, but these can also be specified in a tuning grid as describe above.

Linear regression and discriminant analysis

Linear regression seeks to identify (fit) the hyperplane in multivariate space that minimises the difference between the observed target variable and that predicted by the model. The linear nature of the model relates to this and can be applied for prediction and inference. Discriminant function analysis (Fisher, 1936) can be used to predict membership or class for a discrete group of classes. It extracts discriminant functions from independent variables which are used to generate class membership probabilities for each observation. If there are k groups, the aim is to extract k discriminant functions. An observation is assigned to a group if the value for the discriminant function for the group is the smallest. It was extended from the linear classifier to the quadratic case by Marks and Dunn (1974). Because of the partitioning of the space in this way for clusters, any collinear variables in the data must be removed and any numerical variables converted to factors. Here the lda implementation in caret is used, but you should note that there are many flavours of discriminant analysis included in caret (see https://topepo.github.io/caret/train-models-by-tag.html#discriminant-analysis).

k-nearest neighbour

The kNN algorithm operates under the assumption that records with similar values of, for example, Price in the properties data have similar attributes – they are expected to be close in the multidimensional feature space described earlier. In this sense this algorithm seeks to model continuous variables or classes, given other nearby values. The algorithm is relatively straightforward: for each y_i, select the k observations closest to observation i and predict y_i to be the arithmetic mean of y of these observations. In this example, it is used to compute the average of the k nearest observations. Thus the kNN algorithm is based on feature similarity to the number of neighbours (k) in attribute rather than geographical space.

Bagged regression trees

Bootstrap aggregating or bagging (Breiman, 1996) seeks to overcome the high variance in regression trees by generating multiple models with the same parameters. The variance issue with regression trees is that although the initial partitions at the top of the tree will be similar between runs and sampled data, there can be large differences in the branches lower down between individual trees. This is because later (deeper) nodes tend to overfit to specific sample data attributes in order to further partition the data. As a result, samples that are only slightly different can result in variable models and differences in predicted values. This high variance causes model instability. The models (and their predictions) are also sensitive to the initial training data sample, and thus regression trees suffer from poor predictive accuracy.

To overcome this, bootstrap aggregating (bagging) was proposed by Breiman (1996) to improve regression tree performance. Bagging, as the name suggests, generates multiple models with the same parameters and averages the results from multiple tress. This reduces the chance of overfitting as might arise with a single model and improves model prediction. The bootstrap sample in bagging will on average contain 63% of the training data, with about 37% left out of the bootstrapped sample. This is the out-of-bag sample which is used to determine the model's accuracy through a cross-validation process. There are three steps to bagging:

1. Create a number of samples from the training data. These are termed *bootstrapped* samples because they are repeatedly sampled from a training set, before the model is computed from them. These samples contain slightly different data but with the same distribution properties of the full dataset.

2. For each bootstrap sample create (train) a regression tree.

3. Determine the average predictions from each tree to generate an overall average predicted value.

Random forests

Bagged regression trees may still suffer from high variance, with associated reduced predictive power. They may exhibit tree correlation as the trees generated by the bagging process may not be completely independent of each other due to all predictor variables being considered at every split of every tree. As a result, trees from different bootstrap samples may have a similar structure to each other, and this can prevent bagging from optimally reducing the variance of the model prediction and performance. Random forests (Breiman, 2001) seek to reduce tree correlation. They build large collections of *decorrelated* trees by adding randomness to the tree construction process. They do this by using a bootstrap process and by split-variable randomisation. Bootstrapping is similar to bagging, but repeatedly resamples from the original sample. Trees that are grown from a bootstrap resampled dataset are more likely to be decorrelated. The split-variable randomisation introduces random splits and noise to the response, limiting the search for the split variable to a random subset of the variables. The basic regression random forest algorithm proceeds as follows:

```
 1.  Select the number of trees ('ntrees')
 2.  for i in 'ntrees' do
 3.  | Create a bootstrap sample of the original data
 4.  | Grow a regression tree to the bootstrapped data
 5.  | for each split do
 6.  | | Randomly select 'm' variables from 'p' possible variables
 7.  | | Identify the best variable/split among 'm'
 8.  | | Split into two child nodes
 9.  | end
10.  end
```

Gradient boosting machine

Boosting seeks to convert weak learning trees into strong learning ones, with each tree fitted on a slightly modified version of the original data. Gradient boosting machines (Friedman, 2001) use a loss function to indicate how well a model fits the underlying data. It trains many trees (models) in a gradual, additive and sequential manner, and parameterises subsequent trees by evaluated losses in previous ones (boosting them). The loss function is specific to the model objectives. For the house price models here, the loss function quantifies the error between true and predicted (modelled) house prices. GBMs have some advantages for inference as the results from the boosted regression trees are easily explainable and the importance of the inputs can easily be retrieved.

Support vector machine

Support vector machine (SVM) analyses can also be used for classification and regression (Vapnik, 1995). SVM is a non-parametric approach that relies on kernel functions. For regression, SVM model outputs do not depend on distributions of input variables (unlike ordinary linear regression). Rather SVM uses kernel functions that allow non-linear models to be constructed under the principle of maximal margin, which focuses on keeping the error below a certain value rather than prediction. The kernel functions undertake complex data transformations and then determine the process (hyperplanes) that separates the transformed data. The support vectors are the data points that are closer to the hyperplane, influencing its position and orientation. These support vectors allow the margins to be maximised and removing support vectors changes the position of the hyperplane, allowing the SVM to be built.

7.5.3 Prediction

The code below creates six prediction models using standard linear regression (SLR), k-nearest neighbour (kNN), bagged regression trees (BRTs), random forests (RF), gradient boosting machines (GBMs) and support vector machine (SVM). Each of the models is constructed with the rescaled training data (`train_z`), the same control parameters (a 10-fold cross-validated model) and the default model tuning parameters. The predictive models are applied to the normalised test data (`test_z`) to generate predictions of known house price values which are then evaluated. Finally, the predictions from the different models are compared.

The following control parameters are used in all models:

```
# the control parameters
ctrl <- trainControl(method = "repeatedcv", number = 10, repeats = 10)
```

The code below creates each of the models, with some of them (e.g. the random forest) taking some time to run:

```
lmFit   = train(Price~.,data = train_z, method = "lm",
                trControl = ctrl, verbose = FALSE)
knnFit = train(Price~.,data = train_z, method = "knn",
                trControl = ctrl, verbose = FALSE)
tbFit   = train(Price~.,data = train_z, method = "treebag",
                trControl = ctrl, verbose = FALSE)
rfFit   = train(Price~.,data = train_z, method = "rf",
                trControl = ctrl, verbose = FALSE)
gbmFit  = train(Price~.,data = train_z, method = "gbm",
                trControl = ctrl, verbose = FALSE)
svmFit = train(Price~.,data = train_z,method = "svmLinear",
                trControl = ctrl, verbose = FALSE)
```

Remember that you can save R objects to allow you to come back to these without running them again:

```
# the models
save(list = c("lmFit","knnFit","tbFit","rfFit","gbmFit","svmFit"),
    file = "all_pred_fits.RData")
```

For all of the models, the outputs can be examined in the same way (with a model named xxFit):

- Tuning parameters can be examined using modelLookup.
- The model results can be examined through xxFit$results.
- The best fitted model is reported in zzFit$finalModel.
- Predictions can be made using the predict function, which automatically takes the best model: pred.xx = predict(xxFit, newdata = test_z).
- The prediction can then be evaluated using the postResample function to compare the known test values of Price with the predicted ones: postResample(pred = pred.xx, obs = test_z$Price).

The predicted values for each individual model can be plotted against the observed values as in Figure 7.3.

The effectiveness of model prediction can also be compared, using the postResample function in caret with the generic predict function. Table 7.1 shows the accuracy of the predicted house price values when compared against the known, observed house price values in the test data for each model. Here, it looks like the random forest approach is generating the model with the strongest fit.

```
# generate the predictions for each model
pred.lm = postResample(pred = predict(lmFit, newdata = est_z),
                       obs = test_z$Price)
pred.knn = postResample(pred = predict(knnFit, newdata = test_z),
                        obs = test_z$Price)
pred.tb = postResample(pred = predict(tbFit, newdata = test_z),
                       obs = test_z$Price)
pred.rf = postResample(pred = predict(rfFit, newdata = test_z),
                       obs = test_z$Price)
pred.gbm = postResample(pred = predict(gbmFit, newdata = test_z),
                        obs = test_z$Price)
pred.svm = postResample(pred = predict(svmFit, newdata = test_z),
                        obs = test_z$Price)
# Extract the model validations
df_tab = rbind(
        lmFit$results[which.min(lmFit$results$Rsquared),2:4],
        knnFit$results[which.min(knnFit$results$Rsquared),2:4],
        tbFit$results[which.min(tbFit$results$Rsquared),2:4],
        rfFit$results[which.min(rfFit$results$Rsquared),2:4],
        gbmFit$results[which.min(gbmFit$results$Rsquared),5:7],
        svmFit$results[which.min(svmFit$results$Rsquared),2:4])
colnames(df_tab) = paste0("Model ", colnames(df_tab))
# Extract the prediction validations
df_tab2 = t(cbind(pred.lm, pred.knn, pred.tb,
                  pred.rf, pred.gbm, pred.svm))
colnames(df_tab2) = paste0("Prediction ", colnames(df_tab2))
# Combine
df_tab = data.frame(df_tab, df_tab2)
rownames(df_tab) = c("SLR", "kNN","BRT", "RF", "GBM", "SVM")
# Print out
df_tab
```

Table 7.1   Model and prediction accuracies of different machine learning models

	Model			Prediction		
	RMSE	R^2	MAE	RMSE	R^2	MAE
SLR	110.468	0.570	65.045	88.559	0.627	58.578
kNN	112.959	0.561	61.649	89.702	0.626	49.989
BRT	101.456	0.635	61.100	86.750	0.640	55.006
RF	106.436	0.651	58.858	73.734	0.743	42.426
GBM	113.637	0.558	64.553	77.106	0.717	46.526
SVM	116.153	0.558	58.189	92.950	0.633	52.279

In summary, there are a large number of possible refinements and choices in the models used for prediction:

- Different models will be better suited to the input data than others, suggesting the need to explore a number of model types and families.

- The model predictions can be evaluated by the degree to which they predict observed, known values in the test data.

- The model algorithms can be tuned beyond the `caret` defaults through a tuning grid, and details of the tuning parameters are listed using the `modelLookup("<method>")` function.

- The models could be run outside of `caret` to improve their speed and tuning options as some take a long time to run with even modest data dimensions.

You may wish to save the results of this and the preceding sections (e.g. `save. image(file = "section 7.5.3.RData")`).

7.5.4 Inference

In contrast to *prediction*, where the aim is to create a model able to reliably predict the response variable, given a set of input variables, the aim of *inference* is understanding, specifically, *process* understanding.

The code below defines X and Y to illustrate a different `caret` syntax used in the inferential models. It drops one of the life-stage/age variables (u25) because groups of variables adding to 1, 100, and so on across all records can confound statistical models. It also converts the logical true or false values to characters and the Beds variable from an ordered factor to numeric values:

```
X = data_anal %>% select(-Price, -u25) %>%
    mutate_if(is_logical,as.character) %>%
    mutate(Beds = as.numeric(Beds)) %>%
    mutate_if(is_double,scale) %>% data.frame()
Y = data_anal["Price"]
```

The data considerations for inference are slightly different than for prediction because the aims are different. First, data do not have to be split into training and validation subsets. In fact they should not be split as there is danger that some of the potential for understanding will be lost through the split, and there is no need to hold data back for training and validation. This means that the full structure of the data can be exploited by the model. Second, there is a critical need to consider variable selection (also known as model selection), particularly in the context of collinearity. Model selection is an important component of any regression analysis, but more so for understanding than for prediction, as indicated above: simply, if the aim is understanding and two variables are correlated (i.e. essentially have the same relationship with the target variable) then identifying the effects of each predictor variable on the target variable may be difficult. Determining which predictor variables to include in the analysis is not so important for prediction, where the aim is simply to identify the model with the strongest prediction accuracy.

Collinearity occurs when pairs of predictor variables have a strong positive or negative relationship between each other. Strong collinearity can reduce model reliability and precision and can result in unstable parameter estimates, inflated standard errors and inferential biases (Dormann et al., 2013). As a result, model extrapolation may be erroneous and there may be problems in separating variable effects (Meloun et al., 2002). The risk of collinearity increases as more predictors are introduced and occurs when pairs of predictor variables have a strong positive or negative relationship with each other. It is typically considered a potential problem when data pairs have correlations of less than –0.8 or greater than +0.8 (Comber and Harris, 2018). Collinearity can be handled by variable reduction and transformation techniques such as principal components analysis (PCA) regression or by related approaches such as partial least squares regression (e.g. Frank and Friedman, 1993). Other approaches for variable selection and handling collinearity include penalised approaches such as the elastic net, a hybrid of ridge regression and the lasso – the least absolute shrinkage and selection operator (Zou and Hastie, 2005). The key point is that the failure to correctly specify a model when collinearity is present can result in a loss of precision and power in the coefficient estimates, resulting in poor inference and process understanding.

We can investigate collinearity among the variables in a number of ways. The simplest is to construct a linear regression model and then examine the variance inflation factors (VIFs) using the vif function in the car package. Heuristics for interpreting these are taken from Belsley et al. (2005) and O'Brien (2007): collinearity is likely to be a potential problem with VIFs greater than 10 for a given predictor. The code below constructs a regression model and then returns the input variables with a VIF greater than 2. The results suggest that only very weak collinearity exists within these data:

```
library(car)
m = lm(Price~.,data=cbind(X,Y))
vif(m)[ vif(m) > 2]
```

```
##      Garden    Terraced    Detached    Semi.Detached
##    2.095882    2.093380    4.556362         2.334352
##     Easting         u16         u45              u65
##    2.733526    2.316360    2.371342         2.105384
##         o65
##    2.357073
```

The code below creates six inference models using standard linear regression (SLR), *k*-nearest neighbour (*k*NN), bagged regression trees (BRTs), random forests (RF), gradient boosting machines (GBMs) and support vector machine (SVM). None of these are tuned beyond the defaults in caret. However, tuning has been heavily illustrated in previous sections. The aim here is to illustrate the different types of models, how they can be compared and evaluated and then used to generate understanding of the factors associated with house price. The code below creates each of the models, with the RF taking some time to run:

```
ctrl= trainControl(method="repeatedcv",number=10,repeats=5)
set.seed(123)
lmFit   = train(Price~.,data = cbind(X,Y),method = "lm",
                trControl = ctrl)
knnFit = train(Price~.,data = cbind(X,Y),method = "knn",
                trControl = ctrl,trace = F)
tbFit   = train(Price~.,data = cbind(X,Y),method = "treebag",
                trControl = ctrl)
rfFit   = train(Price~.,data = cbind(X,Y),method = "rf",
                trControl = ctrl,verbose = F,importance = T)
gbmFit = train(Price~.,data = cbind(X,Y),method = "gbm",
                trControl = ctrl,verbose = F)
svmFit = train(Price~.,data = cbind(X,Y),method = "svmLinear",
                trControl = ctrl,verbose = F)
```

For inference and understanding, the aim is to understand how each of the different inputs are related to Price, the target variable. However, each approach generates different kinds of results. For example, coefficient estimates can be extracted from standard regression models with lm (or biglm introduced in Chapter 6), but not from the others.

Examining the *variable importance* allows the different model inferences to be explored. Figure 7.4 compares these for these different models. Each model defines variable importance in different ways, and so the specific mechanism for each model should be examined outside of caret. In this case the results show that the Beds covariate is generally the variable that is associated most strongly with the response (Price), but some models identify others as well: greenspace area (gs_area) for *k*NN and SVM, Northing (all models), and the proportion of the population under 45 (u45), unemployed (unmplyd) and detached properties (Detached) are also generally important. Notice also the different profiles of the variables in the different models: the bagged regression tree (BRT) in Figure 7.4 has many similarly salient variables, whereas the other models have similar gradients/profiles of importance (although with different variables).

```
# define a print function
# this was modified from here:
# https://github.com/topepo/caret/blob/master/pkg/caret/R/print.
varImp.train.R
print.varImp.10 <- function(x = vimp, top = 10) {
    printObj <- data.frame(as.matrix(sortImp(x, top)))
    printObj$name = rownames(printObj)
    printObj
}
# use this to extract the top 10 variables to a data.frame - df
df = data.frame(print.varImp.10(varImp(lmFit)),
                method = "SLR")
df = rbind(df, data.frame(print.varImp.10(varImp(knnFit)),
                method = "kNN"))
df = rbind(df, data.frame(print.varImp.10(varImp(tbFit)),
                method = "BRT"))
df = rbind(df, data.frame(print.varImp.10(varImp(rfFit)),
                method = "RF"))
df = rbind(df, data.frame(print.varImp.10(varImp(gbmFit)),
                method = "GBM"))
df = rbind(df, data.frame(print.varImp.10(varImp(svmFit)),
                method = "SVM"))

df %>%
  ggplot(aes(reorder(name, Overall), Overall)) +
  geom_col(fill = "tomato") +
  facet_wrap( ~ method, ncol = 3, scales = "fixed") +
  coord_flip() + xlab("") + ylab("Variable Importance") +
  theme(axis.text.y = element_text(size = 6)) + theme_bw()
```

There is much more that could be done here. The data were not scaled, none of the approaches was tuned, in-sample accuracies have not been reported, tune lengths could be increased and alternative controls could be evaluated. Changes to all of these default settings could be explored. It is worth noting that to examine any specific model in more detail may require stepping out of the caret implementation to the original package implementation, in order to explore how to exercise greater control and analytical nuance. The source packages for each model are listed at https://topepo.github.io/caret/available-models.html.

Again, you may wish to save the results of this section (e.g. save.image(file = "section 7.5.4.RData")).

7.5.5 Summary of key points

The caret package provides a consistent environment for and integrated interface to many machine learning algorithms, samples of which were used in this section.

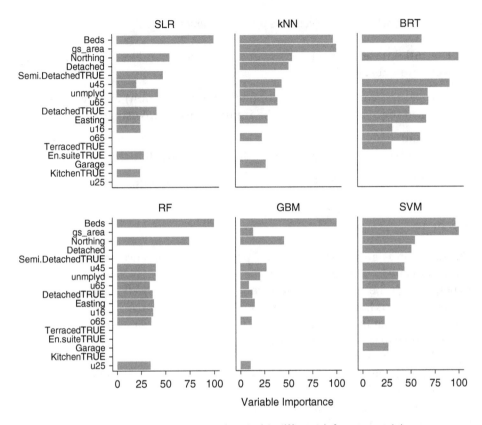

Figure 7.4   The variable importance associated with different inference models

These were illustrated using the Liverpool residential properties data linked to census socio-economic attributes to construct models of house price.

There are two distinct workflows for inference and prediction. For prediction, where the aim is simply to predict a value *as accurately as possible*:

- The data were split into training and validation (testing) subsets.

- The data subsets were rescaled, keeping the target variable in its original form. This should always be done after splitting the data into training and validation subsets.

- Each maximum likelihood approach was applied to the same data using the same control parameters (i.e. they were not extensively tuned as the example in Section 7.4 was) to create a predictive model.

- The generic predict function was used to apply each model to the testing/validation data and the predicted house price values were compared with the observed ones to get measures of model fit.

For inference, where the objective is *process understanding*:

- The full datasets were used to construct models of house price (i.e. they were not subsetted).
- Some of the variables were dropped (model selection) and variable independence/collinearity in the input data was investigated.
- The models were compared through variable importance, which describes the relative explanatory power of different inputs to each model.
- This allows inferences about the processes diving house price to be made.

7.6 CLASSIFICATION

Classification is a different regression process than inference and prediction. In contrast to linear regression, for example, where the objective is to fit some kind of hyperplane in the multivariate feature space that best predicts or describes the outcome (such as house price), the aim in classification is to group or cluster observations, such that the groups have similar properties or characteristics.

There are two broad families of classification approaches: partition-based clustering and hierarchical clustering. Partition-based approaches (such as *k*-means clustering and partitioning around medoids) seek to identify some sort of cluster or class *typicality* in multidimensional feature space, and observations are allocated to the cluster to which they are nearest. Hierarchical clustering seeks to identify groups of observations based on their *similarity*, either grouping them all together before iteratively splitting (divisive clustering) or iteratively grouping individual observations (agglomerative clustering). The key considerations are that:

- allocation and grouping are based on some measure of similarity or multivariate distance;
- partitioning requires k, the number of classes, to be defined in advance;
- in hierarchical clustering the number of clusters can be chosen *post hoc*;
- as with prediction, classification may be *supervised* where the clusters are known in advance, and *unsupervised* where they are not.

You may wish to clear your workspace and reload the data for this chapter:

```
rm(list = ls())
load("ch7.RData")
```

In an *unsupervised* classification the result is a set of typical values in each variable for each class called the *class centres*. Each observation is allocated to the class centre to which it is nearest (in multivariate space). For *supervised* classification, the class centres are defined in advance and then observations are allocated to their nearest class. In unsupervised approaches, where the clusters are defined from the input data, potential clusters are evaluated by the extent to which they minimise the within-cluster variation – the within-cluster sum of the squared multivariate distances between each observation and the class centre to which it is allocated.

Supervised and unsupervised approaches and their considerations are introduced in the subsections below before a full unsupervised example is worked up, comparing different clustering approaches.

7.6.1 Supervised classification

In supervised classification, a similar sequence of operations is undertaken to supervised prediction: a training validation split of the data is defined, the algorithm is trained and then applied to the test data. Data can be normalised and tuning parameters explored. The major difference is the accuracy measures. These are now based on cross-tabulations or contingency tables of predicted against observed class, from which different overall and marginal measures of accuracy can be generated. Essentially the Accuracy value as reported in the caret models reports the proportion of observations correctly predicted and the Kappa includes information on the marginal (per class) accuracies as well. Probably the most comprehensive description of these measures can be found in the remote sensing literature, and Congalton and Green (2002) provide an excellent introduction. Unsupervised classifications require a slightly different set of considerations, principal of which is determining the number of classes.

The OA data include an OAC class label in the OAC attribute and variables that summarise life-stage, the local environment and socio-economic characteristics of the area. We can use these to construct supervised classification models (i.e. by using a training/validation data split). Recall that the OAC classification (Vickers and Rees, 2007; Gale et al., 2016) was created using census area socio-economic attributes, a common approach for geo-demographic classifications (Harris et al., 2005). The models predicting OAC class are constructed using their implementation in the caret package and then validated using the test data.

The code snippets below convert the OAC attribute to a factor, partition the data into training and testing subsets, and rescale the attributes to z-scores (i.e. with a mean of 0 and a standard deviation of 1). The training/validation split of OAC classes is shown in Table 7.2.

```
# convert to factor
oa$OAC = as.factor(oa$OAC)
set.seed(1234)
```

```
# partition the data
train.index <-
  as.vector(createDataPartition(oa$OAC, p=0.75, list = FALSE))
# Create Training Data
train_data = oa[train.index,-1] %>% st_drop_geometry() %>%
  mutate_if(is_double,scale)
test_data = oa[-train.index,-1] %>% st_drop_geometry() %>%
  mutate_if(is_double,scale)
cbind(Train = summary(train_data$OAC),Test = summary
                      (test_data$OAC))
```

Table 7.2   Training/validation split across OAC classes

	Train	Test
Constrained City Dwellers	219	72
Cosmopolitans	125	41
Ethnicity Central	81	26
Hard-Pressed Living	378	125
Multicultural Metropolitans	85	28
Suburbanites	150	50
Urbanites	153	51

Next, a series of predictive models are constructed and evaluated internally, and their accuracy is reported using the accuracy and kappa measures as described above (Table 7.3). Note that again these might take a few minutes to run and you could consider including **verbose** = F for some of them:

```
# set control paramters
ctrl= trainControl(method="repeatedcv",number=10,repeats=10)
# create models
set.seed(123)
ldaMod = train(OAC ~ ., data=train_data, method="lda",
               trControl=ctrl, metric="Accuracy")
knnMod = train(OAC ~ ., data=train_data, method="knn",
               trControl=ctrl, metric="Accuracy")
tbMod  = train(OAC ~ ., data=train_data, method="treebag",
               trControl=ctrl, metric="Accuracy")
rfMod  = train(OAC ~ ., data=train_data, method="rf",
               trControl=ctrl, metric="Accuracy")
gbmMod = train(OAC ~ ., data=train_data, method="gbm",
               trControl=ctrl, metric="Accuracy", verbose = F)
svmMod = train(OAC ~ ., data=train_data, method="svmLinear",
               trControl=ctrl, metric="Accuracy")
```

```
# evaluate the models
round(rbind(
   LDA = ldaMod$results[which.min(ldaMod$results$Accuracy),
                        c("Accuracy", "Kappa")],
   kNN = knnMod$results[which.min(knnMod$results$Accuracy),
                        c("Accuracy", "Kappa")],
   BRT = tbMod$results[which.min(tbMod$results$Accuracy),
                        c("Accuracy", "Kappa")],
   RF  = rfMod$results[which.min(rfMod$results$Accuracy),
                        c("Accuracy", "Kappa")],
   GBM = gbmMod$results[which.min(gbmMod$results$Accuracy),
                        c("Accuracy", "Kappa")],
   SVM = svmMod$results[which.min(svmMod$results$Accuracy),
                        c("Accuracy", "Kappa")]), 3)
```

Table 7.3 Internal model evaluations

	Accuracy	Kappa
LDA	0.591	0.481
kNN	0.533	0.413
BRT	0.586	0.486
RF	0.596	0.498
GBM	0.581	0.478
SVM	0.618	0.520

The models can then be applied to the test data and evaluated again, by comparing the predicted class with the observed one for each OA as in Table 7.4:

```
# generate predictions
pred.lda = postResample(pred = predict(ldaMod,newdata = test_data),
                        obs = test_data$OAC)
pred.knn = postResample(pred = predict(knnMod,newdata = test_data),
                        obs = test_data$OAC)
pred.tb  = postResample(pred = predict(tbMod,newdata = test_data),
                        obs = test_data$OAC)
pred.rf = postResample(pred = predict(rfMod,newdata = test_data),
                        obs = test_data$OAC)
pred.gbm = postResample(pred = predict(gbmMod,newdata = test_data),
                        obs = test_data$OAC)
pred.svm = postResample(pred = predict(svmMod,newdata = test_data),
                        obs = test_data$OAC)
# print out
round(rbind(LDA = pred.lda, kNN = pred.knn, BRT = pred.tb,
            RF = pred.rf, GBM = pred.gbm, SVM = pred.svm),3)
```

Here we see that the most accurate model is the random forest model.

Table 7.4 Classification accuracies of different models

	Accuracy	Kappa
LDA	0.613	0.510
kNN	0.601	0.502
BRT	0.611	0.520
RF	0.631	0.543
GBM	0.608	0.514
SVM	0.618	0.522

For classification models constructed using caret, each predictor variable has a separate variable importance for each class, except for classification trees, bagged trees and boosted trees (see https://topepo.github.io/caret/variable-importance. html). However, for the other models, the variable importance for each class can be examined, as is shown in Figure 7.5 for the SVM model.

```
data.frame(varImp(svmMod)$importance,
          var=rownames(varImp(svmMod)$importance)) %>%
  pivot_longer(-var) %>%
  ggplot(aes(reorder(var, value), value, fill = var)) +
  geom_col() + coord_flip() +
  scale_fill_brewer(palette="Dark2") +
  labs(x = "", y = "Variable Importance for SVM clusters",
       fill = "Attribute")+
  facet_wrap(~name, ncol = 3, scales = "fixed") +
  theme(axis.text.y = element_text(size = 7),
        legend.position = "none")
```

7.6.2 Unsupervised classification

Unsupervised clustering can be illustrated using the classic k-means algorithm. It makes an initial guess at k cluster centres, allocates observations to their nearest cluster centre, evaluates the result and then tries to improve the cluster centres and allocation. The evaluation is usually through the sum of the squared distance between the data points and the cluster centroids, and the iterative improvement seeks to minimise this in order to determine homogeneous clusters with similar data points allocated to the same cluster.

The code below applies the kmeans function in the stats package to determine $k = 10$ centres. This function is automatically loaded with R, and in this case it classifies the OAs into one of 10 classes:

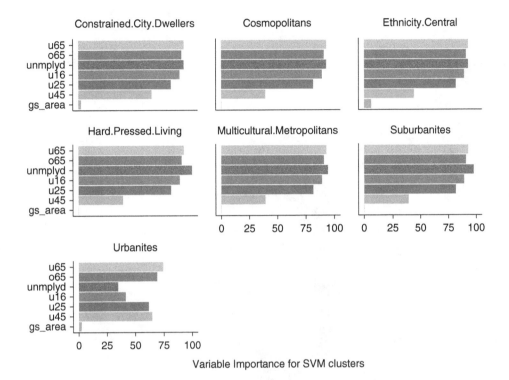

Figure 7.5 Variable importance associated with the SVM classification model

```
set.seed(1234) # reproducibility!
cls <- kmeans(st_drop_geometry(oa[,-c(1, 9)]),
              centers = 10, iter.max = 100, nstart = 20)
```

You should inspect the resulting R object:

```
names(cls)
```

Of interest here are the cluster allocation (cls$cluster), indicating the class to which each observation has been allocated, the within-cluster sum of squares (cls$tot.withinss) that provides a measure of how effective the clusters are relative to the input data, and the class centres (cls$centers). You should inspect these.

The class centres can be attached to the oa data object as an attribute and mapped (Figure 7.6):

```
oa$cls = cls$cluster
# map
```

```
tm_shape(oa) +
  tm_fill("cls", title = "Class No.", style = "cat",
          palette = "Set1")
# remove the 'cls' variable
oa = oa [, -11]
```

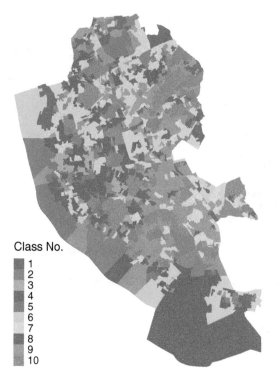

Figure 7.6 Application of a *k*-means classification to the OAs in Liverpool

Unsupervised clustering requires a few more aspects to be considered than unsupervised classification, the main one of which is determining *k*, the number of clusters, and, as an important consideration, the size of the clusters.

In the kmeans function, the evaluation of clusters is through the within-cluster sum of squares (see cls[4] and cls[5]). Thus, by definition, the best partition arises when *k* equals the number of observations. However, this is impractical and *k* can be determined using a number of approaches, the most common of which is the elbow method. This uses the within-cluster sums of squares (WCSS) for different values of *k*. The code below does this for cluster sizes from 1 to 100 and extracts the smallest cluster size (i.e. the number of observations/census areas). These are put into the wss and smallest.clus R objects, respectively. Note that this takes time, so progress is printed out for each 10%:

```
set.seed(1234) # Reproducible outcome
smallest.clus <- wss <- rep(0, 100) # define 2 results vectors
for (i in 1:100) {
    clus <- kmeans(st_drop_geometry(oa[,-c(1, 9)]),
                       centers = i, iter.max = 100, nstart = 20)
    wss[i] <- clus$tot.withinss
    smallest.clus[i] <- min(clus$size)
    if (i%% 10 == 0) cat("progress:", i, "% \n")
}
```

The WCSS values can be plotted against the cluster sizes as a useful diagnostic exercise (as in Figure 7.7). This shows that as the number of clusters increases, the within-cluster variance decreases, as you would expect – each group becomes more homogeneous as the number of groups increases. The plot of the smallest cluster size against the total number of clusters shows the tipping point at which cluster size starts to drop off – again indicating very small, niche and not very universal clusters. Here, it seems that a value of k around 8 has a good position on the elbow of the scree plot and results in a reasonable smallest cluster size:

```
par(mfrow = c(1,2)) # set plot parameters
# the WCSS scree plot
plot(1:100, wss[1:100], type = "h", xlab = "Number of Clusters",
     ylab = "Within Cluster Sum of Squares")
# the smallest cluster size
plot(1:100, smallest.clus[1:100], type = "h",
     xlab = "Number of Clusters",
     ylab = "Smallest Cluster Size")
par(mfrow = c(1,1)) # reset the plot parameters
```

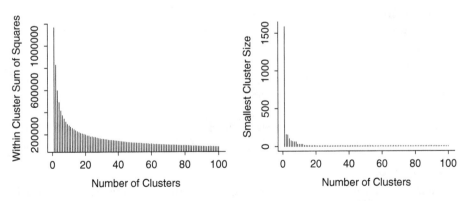

Figure 7.7   Cluster scree plot (left) and smallest cluster size plot (right)

The classification for eight clusters is recalculated – this quantity is stored in n.clus in the code below. This is used in the code below as it provides a reasonable trade-off between WCSS and minimum cluster size and the results can be mapped:

```
set.seed(1234) # reproducibility!
cls <- kmeans(st_drop_geometry(oa[,-c(1, 9)]),
              centers = 8, iter.max = 100, nstart = 20)
oa$cls = cls$cluster
# map
tm_shape(oa) +
   tm_fill("cls", title = "Class No.", style = "cat",
           palette = "Set1")
# remove the 'cls' variable
oa = oa [, -11]
```

7.6.3 Other considerations

There are a number of further considerations in classification.

The first of these relates to alternative methods for determining *k*. The elbow method above is quite subjective and inherently depends on how similarities are measured. One alternative is the average silhouette method, which is similar to the elbow method except that it calculates the average silhouette (i.e. how well each object belongs to its cluster), for different values of *k*, with the aim of finding the maximum average value to determine the appropriate number of clusters. The following code applies this, and indicates 2 as an optimal number of clusters:

```
library(cluster)
set.seed(1234) # Reproducible outcome
smallest.clus <- sil <- rep(0, 99) # define 2 results vectors
for (i in 2:100) {
    clus <- kmeans(st_drop_geometry(oa[,-c(1, 9)]),
                   centers = i, iter.max = 100, nstart = 20)
    ss = silhouette(clus$cluster,
                    dist(st_drop_geometry(oa[,-c(1, 9)])))
    sil[i] = mean(ss[,3])
    smallest.clus[i] <- min(clus$size)
    if (i%% 10 == 0) cat("progress:", i, "% \n")
}
# the WCSS scree plot
plot(2:100, sil[1:99], type = "h",
     xlab = "Number of Clusters",
     ylab = "Average Silhouette")
```

An alternative approach for determining the *k* is the gap statistic. This compares the total within-cluster variation for different values of *k* against an *expected value* when compared to reference data with no clustering, generated through Monte Carlo simulation and sampling. Because of the Monte Carlo/bootstrapping, this can take some time to run and so the evaluation is between 1 and 15 clusters in the code below. The results show that, when clusters are evaluated against values derived from an expected reference distribution (as defined by generating many versions of the reference dataset and taking the average), 5 is the optimal value of k. This is a very different evaluation of the clustering structure in relation to the distribution and structure within the input data.

```
gs <- clusGap(st_drop_geometry(oa[,-c(1, 9)]),
              kmeans, nstart = 25, K.max = 15, B = 50)
df = data.frame(gs$Tab)
plot(1:15, df$gap, type = "b",
     xlab = "Number of Clusters",
     ylab = "Gap Statistic")
abline(v = 5, col = "blue", lty = 2)
```

A second issue to be considered is normalising the data. The use of z-scores was described in earlier sections, but an alternative and frequently used approach, especially with a large number of attributes, is to use a PCA. Essentially, what a PCA tries to do is to identify orthogonal components within the multivariate/multidimensional attribute space that summarise the variation in the data attributes. The idea is that some of the PCA components can be used as summaries of the variables, as the inputs into the classification. Typically, three or four components found through a PCA will explain 70–90% of the variation in the data.

```
PCA <- princomp(st_drop_geometry(oa[,-c(1, 9)])cor = T,scores = T)
cumsum(PCA$sdev^2/sum(PCA$sdev^2))

cumsum(PCA$sdev^2/sum(PCA$sdev^2))
```

Here we can see that the first four components account for 82.3% of the variance in the data, which can be used as input to cluster analysis. This is incredibly useful when the input data have many variables and the results can be evaluated and mapped as before:

```
cls <- kmeans(PCA$scores[,1:4],centers=8,iter.max=100,nstart=20)
```

A final consideration, alluded to at the start of this section, is whether to adopt a hierarchical approach to clustering or a partitioning one. The preceding has heavily illustrated k- means as an example of a partitioning clustering approach. There are a number of potential problems with the k-means approach (for a discussion of these, see https://stats.stackexchange.com/questions/133656/how-to-understand-the-drawbacks-of-k-means). The result is that, for example, very unusual observations tend to form very small clusters because of the nature of the clustering operation which seeks to minimise the within-cluster variation. Small groups of similar but outlying observations can contribute highly to this quantity unless they are assigned their own cluster. This sometimes means that increasing the number of clusters tends to create these very small 'outlier' clusters. Hierarchical clustering (whether top-down divisive or bottom-up agglomerative) generates a dendrogram – a tree-based representations of the data – and makes fewer assumptions about the data and allows the user to decide on the number of clusters by cutting the dendrogram at different heights.

There are different functions for hierarchical clustering in R, including the hclust function in the stats package for agglomerative hierarchical clustering and diana in the cluster package for divisive hierarchical clustering. These are illustrated in turn below:

```
## 1. Agglomerative
# dissimilarity matrix
d <- dist(st_drop_geometry(oa[,-c(1, 9)]), method = "euclidean")
# hierarchical clustering
hc <- hclust(d, method = "complete" )
# plot
plot(hc, cex = 0.6, hang = -1, labels = F,
    xlab = "", main = "Dendrogram of Agglomerative")
# identify 8 clusters
rect.hclust(hc , k = 8, border = 2:12)
## 2. Divisive
# hierarchical clustering
hc <- diana(st_drop_geometry(oa[,-c(1, 9)]))
# the amount of clustering structure
hc$dc
# plot dendrogram
pltree(hc, cex = 0.6, hang = -1, labels = F,
    xlab = "", main = "Dendrogram of Divisive")
rect.hclust(hc , k = 8, border = 2:12)
```

The decision over whether to use a hierarchical approach or a partitioning one will depend on a number of things. Divisive approaches such as *k*-means are better able to handle big data but require *k* to be determined in some way, whereas this can be done by interpreting the dendrogram in hierarchical clustering.

7.6.4 Pulling it all together

The final part of this section generates an unsupervised classification linking the LSOA census data and the housing data to generate a property–socio-economic classification of Liverpool. The data are transformed using PCA and the components used to generate the classification.

Consider the LSOA dataset. This includes variables that summarise life-stage, the local environment and the socio-economic situation. We will use this in an initial example to construct a classification.

In this case we are not trying to predict one of the variables such as house price; rather we are trying to identify groups of similar LSOA areas using a cluster analysis. The code below summarises the numeric attributes (Price and Beds) and then the logical attributes of the properties data over the LSOA data. Recall that the LSOAs contain around 1500 people and that there are 298 LSOAs in the Liverpool area, compared with 1584 OAs:

```
# Part 1: numeric summaries over LSOAs
properties %>% st_transform(27700) %>%
    # intersect with LSOA data and remove geometry
    st_intersection(lsoa) %>% st_drop_geometry() %>%
    # group by LSOAs
    group_by(code) %>%
    # summarise beds and price over LSOAs (areas)
    # mutate beds to numeric
    mutate(Beds = as.numeric(Beds)) %>%
    summarise(BedsM = median(Beds),
              BedsSpread = IQR(Beds),
              PriceM = median(Price),
              PriceSpread = IQR(Price)) -> df1
# Part 2: logical summaries over LSOAs
properties %>% st_transform(27700) %>%
    # intersect with LSOA data
    st_intersection(lsoa) %>%
    group_by(code) %>%
    # summarise logical property values and remove geometry
    summarise_if(is_logical,mean) %>% st_drop_geometry() -> df2
# Part 3: combine with LSOA data
lsoa %>% left_join(df1) %>% left_join(df2) %>%
    st_drop_geometry() %>% dplyr::select(-code) -> class_data
```

Determine the LSOAs with no properties data: these will be removed from the analysis:

```
index.na = which(apply(class_data, 1, function(x) any(is.na(x))))
class_data = class_data[-index.na,]
```

Next, a PCA transformation is applied to the data:

```
PCA <- princomp(class_data, cor = T, scores = T)
PCA$sdev^2/sum(PCA$sdev^2)
```

The code above creates the PCA and lists the proportion of variance explained by each individual component. The next step is to investigate the cumulative amount of variance explained. The aim here is to determine the number of components explaining 70–90% of the variation in the data:

```
cumsum(PCA$sdev^2/sum(PCA$sdev^2))
##     Comp.1     Comp.2     Comp.3     Comp.4     Comp.5     Comp.6
## 0.1432658 0.2388074 0.2882204 0.3330168 0.3700920 0.4041500
##     Comp.7     Comp.8     Comp.9     Comp.10    Comp.11    Comp.12
## 0.4354055 0.4648313 0.4937707 0.5201652 0.5448618 0.5692588
##     Comp.13    Comp.14    Comp.15    Comp.16    Comp.17    Comp.18
## 0.5929655 0.6160866 0.6386988 0.6599927 0.6803223 0.6997149
##     Comp.19    Comp.20    Comp.21    Comp.22    Comp.23    Comp.24
## 0.7185753 0.7364051 0.7537062 0.7705425 0.7870188 0.8030244
##     Comp.25    Comp.26    Comp.27    Comp.28    Comp.29    Comp.30
## 0.8177391 0.8318939 0.8453906 0.8581611 0.8708262 0.8825205
##     Comp.31    Comp.32    Comp.33    Comp.34    Comp.35    Comp.36
## 0.8940712 0.9048314 0.9154465 0.9252348 0.9333868 0.9410854
##     Comp.37    Comp.38    Comp.39    Comp.40    Comp.41    Comp.42
## 0.9486426 0.9556952 0.9623954 0.9688771 0.9749741 0.9807231
##     Comp.43    Comp.44    Comp.45    Comp.46    Comp.47    Comp.48
## 0.9853435 0.9894028 0.9934334 0.9961548 0.9987706 1.0000000
##     Comp.49
## 1.0000000
```

Here we can see that the first 24 components account for 80.3% of the variance in the data. We will use these for a *k*-means cluster analysis.

Again different cluster sizes are evaluated:

```
set.seed(1234) # Reproducible outcome
smallest.clus <- wss <- rep(0, 100) # define 2 variables
for (i in 1:100) {
    clus <- kmeans(PCA$scores[, 1:24],
                    centers = i, iter.max = 100, nstart = 20)
  wss[i] <- clus$tot.withinss
  smallest.clus[i] <- min(clus$size)
  if (i%% 10 == 0) cat("progress:", i, "% \n")
}
```

We can now plot the sums of squares and the cluster sizes as before, with the results suggesting that 10 clusters are appropriate, providing a reasonable trade-off between the WCSS and minimum cluster size:

```
plot(1:100, wss[1:100], type = "h", main = "Cluster Scree Plot",
     xlab = "Number of Clusters",
     ylab = "Within Cluster Sum of Squares")
plot(1:100, smallest.clus[1:100], type = "h", main = "Smallest
Cluster Plot",
     xlab = "Number of Clusters",
     ylab = "Smallest Cluster Size")
set.seed(1234)
clus <- keans(PCA$scores[, 1:24],
              centers = 10, iter.max = 100, nstart = 20)
LSOAclusters = clus$cluster
```

Now the characteristics of the clusters found above can be identified. Since the above code applied the analysis to principal components it is helpful to characterise the clusters in terms of the original variables used in the PCA. To do this, first the z-scores of each variable for the cluster centroids are computed (recall that they have a mean of 0 and a standard deviation of 1). Here, the z values are actually clusterwise mean values, whereas $\bar{x}$ and s are computed for all LSOAs. This is useful for identifying which clusters have relatively high or low values of particular variables. The following code creates a set of z-scores for each cluster:

```
# We need this for the 'ddply' function
library(plyr)
# Compute a data frame (one row per cluster) containing the means
# of each variable in that cluster
mean_by_cluster <-
    ddply(class_data, .(LSOAclusters), numcolwise(mean))[, -1]
# Compute the columnwise means for *all* observations
mean_by_col <- apply(class_data, 2, mean)
# Compute the columnwise *sample* sd's for *all* observations
sd_by_col <- apply(class_data, 2, sd)
# Create the z-scores via the 'scale' function
z_scores <- scale(mean_by_cluster,
                  center = mean_by_col, scale = sd_by_col)
```

The above code used the ddply function in the plyr package. This is used to split the original input data (the class_data data frame) by the LSOAclusters variable which indicates the cluster membership of each census area. This is an example of a split–apply–combine operation. The data frame is split into a number of sub-frames according to the value of LSOAclusters, then an operation is applied to each sub-frame (in this case computing a list of column means – the expression numcolwise(mean) takes a function (mean) and creates a new function, which is then applied to each numeric column in the input data), and finally these are combined together to create a single data frame.

The result can be explored visually. The `heatmap` function provides a visual representation of the `z_scores` variable (Figure 7.8). It generates a grid-based image of the values in the two-dimensional array. However, note that the column and row labels are both categorical, non-ordinal variables. Although the cluster numbers are integers, no significance is attached to their values – they merely serve as labels. This implies that both rows and columns may be permutated, without

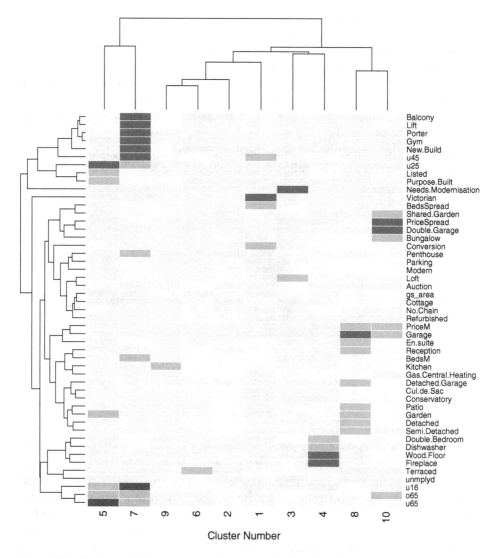

Figure 7.8 Heatmap of the property–socio-economic classification

loss of generality. Doing this can sometimes aid the clarity of the visualisation. In this case, rows are reordered on the basis of applying a dendrogram ordering algorithm to the result of a hierarchical clustering algorithm, where the values along each row are treated as a vector. The clustering is carried out on the basis of Euclidean distance between the vectors using the complete linkage approach (Sørensen, 1948), and the reordering is based on the reorder.dendrogram function provided in the R stats package provided in standard distributions of R. As noted, this reordering is applied to the rows, but following this, a reordering is also applied to the columns – this time treating the columns as vectors, and computing Euclidean distances between these.

```
library(RColorBrewer)
heatmap(t(z_scores),
        scale = 'none',
        col=brewer.pal(6,'BrBG'),
        breaks=c(-1e10,-2,-1,0,1,2,+1e10),
        xlab='Cluster Number', cexRow = 0.6,
        add.expr=abline(h=(0:49)+0.5,v=(0:10)+0.5,col='white'))
```

The shading of the elements of the heatmap is based on the Brewer BrBG palette (see, for example, http://colorbrewer2.org) with category cut-off points at –2, –1, 0, 1, and 2. Thin white lines are added to the grid so that the divisions between individual rows and columns can be seen using the abline function.

Finally, the dendrograms are actually drawn against the axes, so that hierarchical groups of the k-means clusters, and the structure of variables, can be seen. The former is helpful in providing a hierarchical higher-level grouping of clusters, based on their similarities. It is also a useful detector for 'clusters of outliers' as described earlier. For example, the cluster dendrogram in Figure 7.8 suggests that cluster 3 is unusual (note that, hierarchically, it is split away into a one-element group before any other divisions are encountered), and a similar observation can be made for cluster 7.

7.6.5 Summary

There is much more depth to clustering than can be covered in this section. However, the main considerations and concepts have been introduced and illustrated with coded examples. In practice (and *with* practice), key decisions will depend on some understanding of the problem in hand and the objectives of the analysis. Some of the decisions are difficult, such as which approach to use (partitioning or hierarchical). However, the worked example above allows some of the different choices to be explored. It shows how differences between classes can be examined by their typical attributes, and suggests how connected and similar classes are by including a dendrogram of the clusters. It also could be used to examine the impacts of scaling in terms of allocation and class characteristics, by comparing the clusters that are created using different inputs: raw data, data rescaled by z-scores or by PCA. Additional things to consider include the expected distribution of the classes, soft versus hard classification (e.g. fuzzy c-means versus k-means) and the use of central tendencies and assumptions about the distribution of class values around those, rather than, say, Gaussian mixture models which assume a different distribution of data points than methods based around covariance.

Key Points

- The aim in classification is to group or cluster observations, such that the groups have similar properties or characteristics.

- Partition-based clustering seeks to identify some sort of cluster or class *typicality* in multidimensional feature space, with observations allocated to the cluster to which they are nearest.

- Hierarchical clustering seeks to identify groups of observations based on their *similarity*.

- In both cases, clustering is based on some measure of similarity or multivariate distance.

- Partitioning requires k, the number of classes, to be defined in advance.

- In hierarchical clustering the number of clusters can be chosen *post hoc*.

- Classification may be *supervised* where the clusters are known in advance, and *unsupervised* where they are not.

- Supervised partitioning classification was undertaken using the `caret` package and has a similar sequence of operations to supervised prediction.

  - The accuracy measures are derived from a correspondence table comparing the modelled/predicted class against the observed classes in the testing/validation data.

- Unsupervised partitioning classification was illustrated using the classic k-means algorithm applied to all of the data.

  - The clusters are evaluated by a within-cluster sum of squares measure, which provides an indication of the how far each class member is from the cluster centre, for all classes.

  - Determining k, the number of clusters, is a critical consideration in unsupervised approaches (there is no right answer!), and the elbow method is often used, although other approaches exist (such as the average silhouette method and the gap statistic).

- For both approaches, data with many variables (i.e. a large m) can be transformed using a principal components analysis, whose components can be used as inputs.

- A hierarchical clustering approach was illustrated using the `hclust` function.

  ○ This makes fewer assumptions about the data, and the user can decide the number of clusters by examining the dendrogram.

- A worked example was used to create a property–socio-economic classification of Liverpool, and the `heatmap` function was used to generate a visual representation of the classes and their characteristics.

REFERENCES

Belsley, D. A., Kuh, E. and Welsch, R. E. (2005) *Regression Diagnostics: Identifying Influential Data and Sources of Collinearity*. Hoboken, NJ: John Wiley & Sons.

Breiman, L. (1984) *Classification and Regression Trees*. Belmont, CA: Wadsworth International Group.

Breiman, L. (1996) Bagging predictors. *Machine Learning*, 24(2), 123–140.

Breiman, L. (2001) Random forests. *Machine Learning*, 45(1), 5–32.

Brunsdon, C. and Comber, L. (2018) *An Introduction to R for Spatial Analysis and Mapping* (2nd edn). London: Sage.

Comber, A. and Harris, P. (2018) Geographically weighted elastic net logistic regression. *Journal of Geographical Systems*, 20(4), 317–341.

Congalton, R. G. and Green, K. (2002) *Assessing the Accuracy of Remotely Sensed Data: Principles and Practices*. Boca Raton, FL: CRC Press.

Dormann, C. F., Elith, E., Bacher, S., Buchmann, C., Carl, G., Carré, G. et al. (2013) Collinearity: A review of methods to deal with it and a simulation study evaluating their performance. *Ecography*, 36(1), 27–46.

Fisher, R. A. (1936) The use of multiple measurements in taxonomic problems. *Annals of Eugenics*, 7(2), 179–188.

Frank, I. E. and Friedman, J. H. (1993) A statistical view of some chemometrics regression tools. *Technometrics*, 35(2), 109–135.

Friedman, J. H. (2001) Greedy function approximation: A gradient boosting machine. *Annals of Statistics*, 29(5), 1189–1232.

Gale, C. G., Singleton, A., Bates, A. G. and Longley, P. A. (2016) Creating the 2011 Area Classification for Output Areas (2011 OAC). *Journal of Spatial Information Science*, 12, 1–27.

Harris, R., Sleight, P. and Webber, R. (2005) *Geodemographics, GIS and Neighbourhood Targeting*. Chichester: John Wiley & Sons.

James, G., Witten, D., Hastie, T. and Tibshirani, R. (2013) *An Introduction to Statistical Learning*. New York: Springer.

Kuhn, M. (2015) caret: Classification and Regression Training. Astrophysics Source Code Library. https://ascl.net/1505.003.

Kuhn, M. and Johnson, K. (2013) *Applied Predictive Modeling*. New York: Springer.

Marks, S. and Dunn, O. J. (1974) Discriminant functions when covariance matrices are unequal. *Journal of the American Statistical Association*, 69(346), 555–559.

Meloun, M., Militký, J., Hill, M. and Brereton, R. G. (2002) Crucial problems in regression modelling and their solutions. *Analyst*, 127(4), 433–450.

O'Brien, R. M. (2007) A caution regarding rules of thumb for variance inflation factors. *Quality & Quantity*, 41(5), 673–690.

Openshaw, S. (1984) *The Modifiable Areal Unit Problem*, Catmog 38. Norwich: Geo Abstracts.

Sørensen, T. J. (1948) *A Method of Establishing Groups of Equal Amplitude in Plant Sociology Based on Similarity of Species Content and Its Application to Analyses of the Vegetation on Danish Commons*. Copenhagen: I kommission hos E. Munksgaard.

Vapnik, V. (1995) *The Nature of Statistical Learning Theory*. New York: Springer.

Vickers, D. and Rees, P. (2007) Creating the UK National Statistics 2001 output area classification. *Journal of the Royal Statistical Society, Series A*, 170(2), 379–403.

Zou, H. and Hastie, T. (2005) Regularization and variable selection via the elastic net. *Journal of the Royal Statistical Society, Series B*, 67(2), 301–320.

8

ALTERNATIVE SPATIAL
SUMMARIES AND VISUALISATIONS

8.1 OVERVIEW

This chapter describes alternative approaches for visualising spatial data using cartograms, hexagonal binning and tile maps. Alternative visualisation approaches for spatial data are needed because of the so-called *invisibility problem*, in which the characteristics or properties of small but important areas or units may be difficult to pick out or may obscured by other mapped features under classic cartographic representations such as those described in Chapter 5. Cartograms provide a way to overcome the invisibility problem by distorting the reporting units in order to communicate data properties and trends. This chapter illustrates potential solutions to the invisibility problem using a small worked example before a larger example is introduced analysing the ~120 million records in the prescribing dataset, described in earlier chapters.

You will need to have the following packages loaded into your R session:

```
library(maptools)
library(tmap)
library(sf)
library(cartogram)
library(geogrid)
library(GISTools)
library(RSQLite)
library(rpart)
library(visNetwork)
library(tidyverse)
```

8.2 THE INVISIBILITY PROBLEM

Consider the final map in Chapter 4 (Figure 4.3). It shows the spatial variation in opioid prescribing rates in 2018 for the ~32,000 Lower Super Output Areas (LSOAs) in England. However, while some units are distinct, others are hidden in the map due to their small area – the detail of the few areas of high prescribing rates in London is almost entirely lost. This is because the UK census reporting areas were designed to provide a consistent spatial unit for handling *population* geography at different scales. Output Areas (OAs; Martin, 1997, 1998, 2000, 2002) contain around 300 people and LSOAs around 1500 people. This means that their size is a function of population density. This is quite a reasonable basis for reporting information about populations, as opposed to land use for example, and provides a more comparable areal unit than census blocks or tracts in the USA for instance, with 600–3000 and 1000–8000 people, respectively.

One of the consequences of a consistent population-based reporting unit is that when values for those units are mapped, small areas with very large or very small values may be obscured, or if they are visible their importance may be underplayed. Figure 8.1 shows the distribution of the `llti` variable (the proportion of people with limiting long-term illness, LLTI) across the LSOAs in England and Wales. However, what is not displayed is the full picture of LLTI. In this case, then, an area-based geography may be misleading – there are evident problems in the more rural regions in the East and in mid-Wales, but not to the same degree as in the ex-industrial regions in the North/Midlands rust belt (from Liverpool in the North West to Sheffield and Nottingham in the East). Reporting units are smaller in urban settings, with the result that many potentially important patterns are not obvious (LLTI is common in the major post-industrial urban conurbations, with small, dense LSOAs). Here the census geography is occluding information and preventing comparisons across different units. This exemplifies the *problem of invisibility* in choropleth mapping (Harris et al., 2017a), with important patterns obscured especially, not just because the reporting units are small but also because they have varying areal properties.

Load the data from Chapter 4 and then construct the plot, which takes time to compile:

```
load("ch4_db.RData")

# join spatial and attribute data and pipe to tmap
lsoa_sf %>% left_join(social) %>%
    tm_shape() +
    tm_fill("llti", palette = "GnBu",
            style = "kmeans", legend.hist = T) +
    tm_layout(title = "Limiting Long Term Illness",
              frame = F, legend.outside = T,
              legend.hist.width = 1, legend.format = list(digits = 1),
              legend.outside.position = c("left", "top"),
              legend.text.size = 0.7, legend.title.size = 1)
```

To overcome the invisibility problem, areal units can be distorted, converted, or even displayed in a different way, but the problems of invisible dense regions of the map may persist, for example if the data are converted to point symbols. *Cartograms* and *hexbins* provide area-based approaches for changing the geographic representation of areal reporting units while retaining some of the topology (adjacency between areas etc.) and preserving all of the data values.

Key Points

- Areal reporting units are commonly used in choropleth maps to show the spatial distribution of continuous variables.

- However, these are often designed to provide consistent reporting of *population* geography, with the result that smaller units may suffer from the invisibility problem in densely populated regions.

- Cartograms have been proposed as a way of overcoming this problem.

8.3 CARTOGRAMS

Cartograms provide a way of revealing *invisible* places. They break the link between the size of units on the map and their actual physical size. Cartograms rescale units according to some attribute (e.g. llti). There is a relatively long literature on the use of cartograms, with perhaps Tobler (2004) providing the

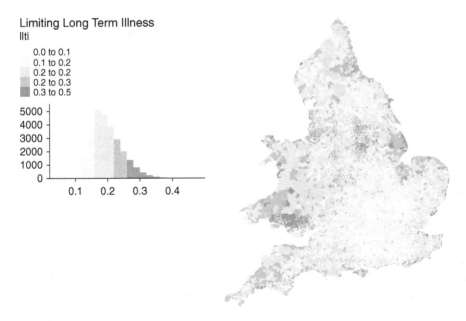

Figure 8.1   Limiting long-term illness for LSOAs in England and Wales

most accessible introduction to their history, but see also the more recent review in Nusrat and Kobourov (2016). There are a number of flavours of cartogram, as will be illustrated, but what they all seek to do is to transform spatial data/areal reporting units by using the value of some property or variable, while maintaining as much as possible of the topology (i.e. the spatial relationships between units) of the individual units and the whole dataset. That is, the geography is changed in order to better describe the spatial distribution of the (usually) area-based features, but the spatial relationships between and relative positioning of individual areas remain consistent.

There are three basic types of cartogram, each of which is illustrated below:

1. Contiguous cartograms

2. Non-contiguous cartograms

3. Dorling cartograms.

These will be explored using the cartogram package and the unemployed variable in the nottingham dataset re-created in the same way as in Chapter 5:

```
# create clip box
ymax = 53.00; ymin = 52.907; xmin = -1.240; xmax = -1.080
pol = st_polygon(list(matrix(c(xmin,ymin,xmax,ymin,
                               xmax,ymax,xmin,ymax,xmin,ymin),
                        ncol=2, byrow=TRUE)))
pol = st_sfc(st_cast(pol, "POLYGON"), crs = 4326)
pol = st_transform(pol, 27700)
# clip out from lsoa_sf and join with social
lsoa_sf[pol,] %>% left_join(social) -> nottingham
# create a map of the original data
p1 = tm_shape(nottingham) +
   tm_polygons("unemployed", palette = "viridis", style = "kmeans") +
   tm_layout(title = "Original", frame = F, legend.show=FALSE)
```

The *contiguous cartogram* is the classic example. It maintains the topology (connectivity) between adjacent areas, but resizes them according to the attribute value of interest. It tries to maintain the original shape of each area, but inevitably resizing comes at a distortion cost. A *key assumption* is that the original areas have some sort of meaning or are familiar to the viewer. Indeed the distortions only have meaning if the viewer is familiar with the original representation and therefore able to interpret the distortion (Figure 8.2). In this case the cartogram package makes use of the rubber sheet distortion algorithm (Dougenik et al., 1985), although the most widely used is probably the density equalising cartogram (Gastner and Newman, 2004).

```
library(cartogram)
# contiguous cartogram
n_cart <- cartogram_cont(nottingham, "unemployed", itermax=10)
p2 = tm_shape(n_cart) +
   tm_polygons("unemployed", palette = "viridis", style = "kmeans") +
   tm_layout(title = "Contiguous", frame = F, legend.show=FALSE)
```

The *non-contiguous cartogram* simply resizes areas according to the value of interest. Each area or polygon is displayed separately, with the loss of adjacency and topology, although this can be inferred by including the original polygon outlines as in Figure 8.2. The cartogram package code uses the non-contiguous area approach of Olson (1976). The advantages of this approach are that original *form* of the map is preserved and the need to be familiar with the original interpretation of the cartogram is reduced. But small areas with large values are only distinguished from their lower-valued neighbours by a reduction in the emphasis placed on those neighbours.

```
n_ncart <- cartogram_ncont(nottingham, "unemployed")
p3 = tm_shape(nottingham) + tm_borders(col = "lightgrey") +
   tm_shape(n_ncart) +
   tm_polygons("unemployed", palette = "viridis", style = "kmeans") +
   tm_layout(title = "Non-contiguous", frame = F, legend.show=FALSE)
```

Dorling cartograms (Dorling, 1996) create non-overlapping symbols that are sized in proportion to the data values (proportional symbols). They use proportional circles, for which a spatial arrangement as close to the original is iteratively sought, preserving some of the topology. This also requires some playing with the cartogram_dorling function parameters to get a cartogram that preserves location (Figure 8.2).

```
# without parameter tweaking!
# n_dorl <- cartogram_dorling(nottingham, "unemployed")
# with parameter tweaking!
n_dorl <- cartogram_dorling(nottingham,
                      "unemployed", k = 0.5, 1, 5000)
p4 = tm_shape(nottingham)+tm_borders(col = "lightgrey")+
   tm_shape(n_dorl)+
   tm_polygons("unemployed", palette = "viridis", style = "kmeans")+
   tm_layout(title = "Dorling", frame = F, legend.show=FALSE)
tmap_arrange(p1,p2,p3,p4, ncol = 2)
```

Key Points

- Cartograms break the link between the mapped size of the reporting units on the map and their actual physical size in reality.

- Areas (units) are rescaled areas according to some attribute, while seeking to preserve as much of the topology as possible (i.e. the spatial relationships between units).

- Three types of cartogram were illustrated: contiguous, non-contiguous and Dorling.

- The critical assumption in the use of cartograms is that the original units have some sort of meaning.

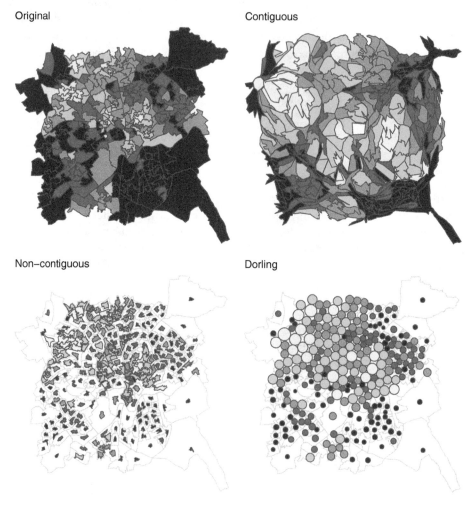

Figure 8.2 Original map of LSOA unemployment rates in Nottingham and the different cartograms

8.4 HEXAGONAL BINNING AND TILE MAPS

Different flavours of cartograms provide a way to overcome the invisibility problem in standard choropleth maps by spatially exaggerating areas with high values and reducing the map space given to areas with low values. However, all cartograms, but especially contiguous cartograms, impose a high level of distortion – the *distortion problem* (Harris et al., 2018), changing the shape and locations of areas on the map, such that areas with the lowest values are both allocated the lowest amount of map space and distorted. They can also massively over- or underemphasise the difference between values at different locations, resulting in different kinds of *misrepresentation* (Harris et al., 2018).

This presents a challenge: cartograms can fail if the level of geographical distortion makes the map difficult to interpret, despite the aim of improving the communication of spatial data (Harris et al., 2017b). There are a number of potential approaches for dealing with misrepresentation. One critical but frequently overlooked solution is to make sure the original map is always presented alongside the cartogram.

Another is to carefully rescale the attribute value that is being mapped in order to achieve an acceptable balance between invisibility and distortion, using square roots or logs of counts (squares or exponential for proportions). Figure 8.3 compares contiguous cartograms scaled in this way.

```
# rescale the value
nottingham$u_sq = (nottingham$unemployed)^3
nottingham$u_exp = exp(nottingham$unemployed)
# create cartograms
n_cart_sq <- cartogram_cont(nottingham, "u_sq", itermax=10)
n_cart_ex <- cartogram_cont(nottingham, "u_exp", itermax=10)
# create tmap plots
p3 =
  tm_shape(n_cart_sq) +
  tm_polygons("u_sq", palette = "viridis", style = "kmeans") +
  tm_layout(title = "Squares", frame = F, legend.show=FALSE)
p4 =
  tm_shape(n_cart_ex) +
  tm_polygons("u_exp", palette = "viridis", style = "kmeans") +
  tm_layout(title = "Exponential", frame = F, legend.show=FALSE)
# plot with original
tmap_arrange(p1, p2, p3, p4, ncol = 2)
```

A further solution is to use equal area tile maps – an equal area cartogram – giving equal space to all map areas and making them all equally visible. Commonly used tile shapes include squares and hexagons, which are shaded in the same way as a normal choropleth. The geogrid package provides functions to do this.

Original

Contiguous

Squares

Exponential

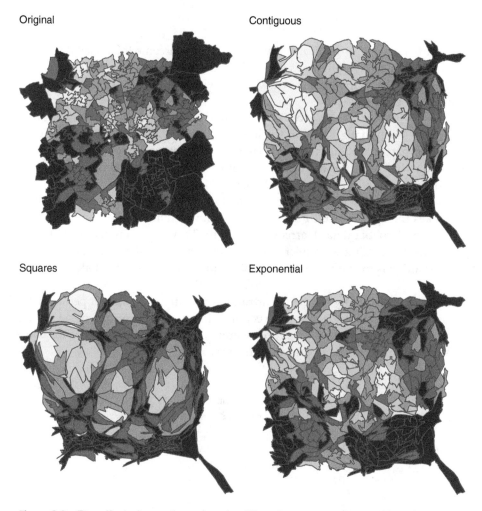

Figure 8.3   The effect of rescaling values in different ways on cartogram distortion

First an initial grid is established of n grid cells, where n is the number of polygons in the input data. The arrangement of these mimics the coverage of the input data:

```
hg = calculate_grid(nottingham, learning_rate = 0.05,
                    grid_type = "hexagonal", verbose = F)
```

The spatial properties of the result can be compared with the input as in Figure 8.4:

```
tm_shape(hg[[2]], bbox = nottingham)+tm_borders() +
  tm_shape(nottingham) + tm_borders(col = "darkgrey") +
    tm_layout(title = "Hexbins", frame = F, legend.show=FALSE)
```

Hexbins

Figure 8.4 The hexagonal grid and the original LSOA data

Then each of the original area/polygons is allocated to the nearest grid. This requires a combinatorial optimisation routine to assign each original area to a new grid cell, in such a way that the distances between the old and new locations are minimised across all locations using Kuhn's *Hungarian method* (Kuhn, 1955). This can take time to solve:

```
hg = assign_polygons(nottingham, hg)
```

And of course the same can be done to create a square grid – again this takes time:

```
sg = calculate_grid(nottingham, learning_rate = 0.05,
                    grid_type = "regular", verbose = F)
sg = assign_polygons(nottingham, sg)
```

Then the results can be plotted as in Figure 8.5:

```
p2 = tm_shape(hg) +
   tm_polygons("unemployed", palette = "viridis", style = "kmeans") +
   tm_layout(title = "Hexagon bins", frame = F, legend.show=FALSE)
p3 = tm_shape(sg) +
   tm_polygons("unemployed", palette = "viridis", style = "kmeans") +
   tm_layout(title = "Square bins", frame = F, legend.show=FALSE)
tmap_arrange(p1, p2,p3, ncol = 3)
```

Original   Hexagon bins   Square bins

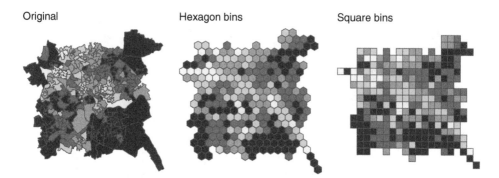

Figure 8.5 Hexagonal and square bin grids showing the spatial distribution of employment alongside the original areas

Key Points

- Cartograms can suffer from distortion problems which over- or underemphasise the differences between values at different locations.

- Such misrepresentations can be overcome by always presenting the original map alongside the cartogram, by carefully rescaling the attribute value that is being mapped or by using equal area tile maps, which use heuristic searches to allocate each original area to a tile.

8.5 SPATIAL BINNING DATA: A SMALL WORKED EXAMPLE

Equal area tiles can also be used as a spatial framework to collate, aggregate and summarise other data – to *bin* data. The basic idea is the same as the use of geom_ hex to summarise the number of data points with similar x and y values, as was done in Chapter 5. But rather than plot them in a scatter plot, with the bin shades indicating the number of data points in each x–y range, here the approach used is to summarise data points falling within certain spatial locations.

The principle can be applied to spatial data points that record the occurrence of something, with longitude and latitude or easting and northing being used instead of x and y.

First, it is instructive to examine a familiar example. Figure 8.6 shows the prop- erties data in Liverpool, introduced in Chapter 7, summarised over hexbins representing LSOAs in Liverpool. The code below loads the properties simple features object of houses for sale in Liverpool. It then uses the lsoa_sf data and the st_make_grid function in sf to create a set of 500 m hexbins covering the study area. Load the Chapter 7 data with the properties layer:

```
load("ch7.RData")
```

Then create and map the hexbins:

```
# transform to OSGB projection
props = st_transform(properties, 27700)
ymax = 53.5066; ymin = 53.32943; xmin = -3.092724; xmax = -2.790667
pol = st_polygon(list(matrix(c(xmin,ymin,xmax,ymin,
                               xmax,ymax,xmin,ymax,xmin,ymin),
                        ncol=2, byrow=TRUE)))
# convert polygon to sf object with correct projection
pol = st_sfc(st_cast(pol, "POLYGON"), crs = 4326)
# re-project it
pol = st_transform(pol, 27700)
# clip out Liverpool from lsoa_sf
liverpool = lsoa_sf[pol,]
# map
lg = st_make_grid(liverpool, 500, what = "polygons", square = F)
tm_shape(lg)+ tm_borders()+ tm_shape(st_geometry(props))+
   tm_dots(col = "#FB6A4A", alpha = 0.5, size = 0.15)
```

Then the number properties in each hexbin can be determined and mapped. The tools for doing this are just easier in sp. The results are shown in Figure 8.7.

```
# convert to sp - the processing of values is just easier!
lg = as(lg, "Spatial")
props = as(props, "Spatial")
lg$house_count = colSums(gContains(lg, props, byid = TRUE))

# the sf code for doing this is below
# lg <- st_as_sf(lg)
# lg$ID = 1:nrow(lg)
# find points within polygons
# intersection <- st_intersection(lg, y = props)
# int_result <- intersection %>% group_by(ID) %>% count() %>%
st_drop_geometry()
# lg <- lg %>% left_join(int_result) %>% rename(house_count = n)
# lg$house_count[is.na(lg$house_count)]<-0

# create a map with an OSM backdrop
tmap_mode("view")
tm_shape(lg)+
   tm_polygons("house_count", title = "Hexbin Counts",
               breaks = c(0, 1, 5, 10, 20, 40),
               palette = "Greens", alpha = 0.5, lwd = 0.2)+
   tm_layout(frame = F, legend.outside = T)+
   tm_basemap('OpenStreetMap')+ tm_view(set.view = 10)
# reset viewing mode
tmap_mode("plot")
```

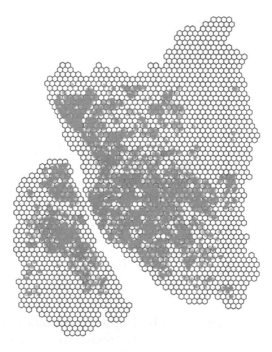

Figure 8.6   Hexagonal bin grid and properties in Liverpool

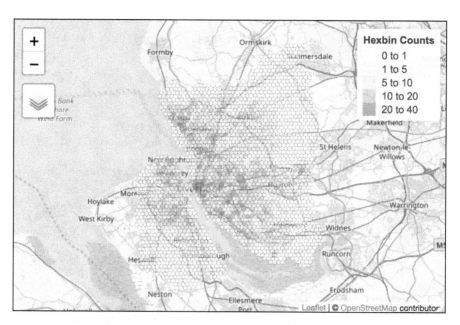

Figure 8.7   A choropleth map of the counts of properties in each hexbin grid, with an OpenStreetMap backdrop

8.6 BINNING LARGE SPATIAL DATASETS: THE GEOGRAPHY OF MISERY

Having established the principle of hexbinning, we can now explore a more complex study to examine the spatial distribution of antidepressants, specifically selective serotonin reuptake inhibitors (SSRIs). These are widely used to treat depression, particularly in persistent or severe cases. This develops the ideas introduced in Chapter 4 in more detail.

8.6.1 Background context

Identifying the geographical patterns of mental ill health ('misery') in relation to underlying socio-economic and especially ethnic factors is useful. It can:

* enhance understanding of the spatial variations in mental ill health in relation to different social factors and the underlying social processes they imply;

* generate evidence to inform judgements regarding best/worst practice in treatment and referral;

* if examined over time (here just 2018 is considered), quantify the impacts of the quasi-natural experiment of austerity policies that started in 2010 in the UK;

* counter overly simplistic narratives around community well-being, for example those based entirely upon personal self-reporting as is current practice (the results of which are largely meaningless).

The research context is that experiences of misery typically attract a range of psychiatric diagnoses, including depression and anxiety disorder. They are increasingly prevalent and generate significant societal, personal and economic costs (estimated £77 billion per year in 2020). They are powerfully influenced by interacting social and economic policies and conditions. Misery is more common among the unemployed and those on low incomes. It typically increases with social inequality, and there is evidence that inequality is causal. Among UK ethnic minorities, mental health problems overall are more common (including diagnoses of depression and, among Black Caribbeans, schizophrenia) yet fewer gain access to services. Gender is also a factor, with women overrepresented in mental health services. They are more likely than men to be given diagnoses of depression or anxiety disorder, while men are more likely to be identified with drug or alcohol problems. Other factors that interact with these include physical health and disability, levels of social support and social isolation, and histories of abuse and neglect. The effects of all these factors can be either mitigated or magnified by policy decisions. However, these possibilities are obscured by current emphases upon well-being as opposed to misery.

Additionally, at the same time, there are embedded tendencies to treat poor mental health as an individual, medical problem rather than link these community-level socio-economic conditions. The code below provides an initial analysis to support an integrated mental ill-health observatory to provide the evidence base that identifies the varying but locally significant factors associated with mental ill health – the geography of misery, as described in Comber et al. (2020).

8.6.2 Extracting from and wrangling with large datasets

Prescription data are held in the `prescribing.sqlite` dataset. This contains ~120 million prescribing records by general practitioners (GPs – doctors) in England for 2018, as described in Chapter 4, as well as other related information about the distribution of patients and socio-economic data.

You should clear your R session of all objects:

```
rm(list = ls())
```

Then connect to the full `prescribing.sqlite` database that you downloaded in Chapter 4. You may need to point your RStudio session to the location of the database using the `setwd` function or by using the menu system. In RStudio this is **Session > Set Working Directory > Choose Directory**. In R it is **Misc > Change Working Directory**.

The code snippets below query the database. Recall from Chapter 4 that we want to keep the very large data tables in the database away from our working memory until they have been summarised. The code extracts records for SSRIs, summarises this for each GP practice, and links the results to the spatial distribution of patients. It is done in a number of discrete steps to aid transparency, and at the end of each of these a `tibble` of the summary information is returned to working memory. Note that each step below undertakes a portion of the analysis, gradually building to a single operation that undertakes it all.

First you need to connect to the database. The code below simply opens the connection to the `prescribing.sqlite` and lists all the tables and fields:

```
# connect to the full database
db <- dbConnect(SQLite(), dbname="prescribing.sqlite")
# list tables and fields
dbListTables(db)
dbListFields(db, "patients")
```

The code below replicates the steps outlined in Chapter 4, which also describes each step individually. In brief, the steps are as follows:

1. Extract the prescriptions of interest, summarising the total number (n) and cost (`tot_cost`) for each GP practice.

2. Determine the distributions of patients across all LSOAs by GP practice.

3. Summarise and link the prescriptions distributions over the LSOAs: the costs, the counts of prescriptions and the prescribing rates.

4. Extract and link to the LSOA socio-economic data from the `social` database table.

The code for undertaking these steps is below:

```
tbl(db, "prescriptions") %>%
  # step 1
  filter(bnf_code %like% '040303%') %>%
  group_by(practice_id) %>%
  summarise(prac_cost = sum(act_cost, na.rm = T),
            prac_items = sum(items, na.rm = T)) %>%
  ungroup() %>%
  # step 2
  left_join(
    tbl(db, "patients") %>%
      group_by(practice_id) %>%
      summarise(prac_pats = sum(count, na.rm = T)) %>%
      ungroup() %>%
      left_join(tbl(db, "patients")) %>%
      mutate(pats_prop = as.numeric(count) / as.numeric(prac_pats))
  ) %>%
  # step 3
  mutate(lsoa_cost = prac_cost*pats_prop,
         lsoa_items = prac_items*pats_prop) %>%
  group_by(lsoa_id) %>%
  summarise(lsoa_tot_cost = sum(lsoa_cost, na.rm = T),
            lsoa_tot_items = sum(lsoa_items, na.rm = T)) %>%
  ungroup() %>%
  filter(!is.na(lsoa_tot_cost)) %>%
  # step 4
  left_join(
    tbl(db, "social") %>%
      mutate(population = as.numeric(population))
  ) %>%
  mutate(cost_pp = lsoa_tot_cost/population,
         items_pp = lsoa_tot_items/population) %>%
  collect() -> lsoa_result
```

The database connection can now be closed and the result examined, after a bit of reordering:

```
dbDisconnect(db)
names(lsoa_result)
```

```
##  [1] "lsoa_id"        "lsoa_tot_cost"  "lsoa_tot_items"
##  [4] "population"     "employed"       "unemployed"
##  [7] "noqual"         "l4qual"         "ptlt15"
## [10] "pt1630"         "ft3148"         "ft49"
## [13] "llti"           "ruc11_code"     "ruc11"
## [16] "oac_code"       "oac"            "cost_pp"
## [19] "items_pp"
```

```
lsoa_result = lsoa_result[, c(1, 2, 3, 18,19, 4:17)]
lsoa_result %>% arrange(-cost_pp)
```

```
## # A tibble: 32,935 x 19
##     lsoa_id lsoa_tot_cost lsoa_tot_items   cost_pp  items_pp
##     <chr>            <dbl>          <dbl>     <dbl>     <dbl>
##  1 E01018...        11003.          9537.     5.27      4.57
##  2 E01017...         4578.          3109.     3.90      2.65
##  3 E01015...         6475.          4955.     3.80      2.91
##  4 E01019...         5600.          3921.     3.61      2.52
##  5 E01015...         3886.          2990.     3.59      2.76
##  6 E01015...         7135.          5475.     3.51      2.69
##  7 E01015...         4372.          3364.     3.32      2.55
##  8 E01023...         4995.          2716.     3.03      1.65
##  9 E01022...         4220.          2532.     2.89      1.73
## 10 E01033...         3312.          2433.     2.89      2.12
## # ... with 32,925 more rows, and 14 more variables:
## #    population <dbl>, employed <dbl>, unemployed <dbl>,
## #    noqual <dbl>, l4qual <dbl>, ptlt15 <dbl>, pt1630 <dbl>,
## #    ft3148 <dbl>, ft49 <dbl>, llti <dbl>, ruc11_code <chr>,
## #    ruc11 <chr>, oac_code <int>, oac <chr>
```

Mapping and aggregating this dataset are described in the next subsections, but at this point you might want to have a look at the data, the relationships between variables, and so on using the approaches described in earlier chapters. As an illustration the code below constructs a regression tree of prescribing rates. Note that this is just an illustration, but it shows the association of different socio-economic variables with llti and how they explain much of the variation in prescribing rates (cost per person) of these antidepressants. This reinforces how useful it is to examine the *data structure* prior to formal data analysis (see Chapter 4).

```
library(rpart)
library(visNetwork)
lsoa_result$oac = as.factor(lsoa_result$oac)
```

```
m1 <- rpart(cost_pp ~
            unemployed+noqual+l4qual+ptlt15+pt1630+ft3148+ft49+llti,
            data = lsoa_result, method = "anova")
visTree(m1,legend=FALSE,collapse=TRUE,direction='LR')
```

8.6.3 Mapping

It would be relatively straightforward to simply map the counts or rates. The code below reloads the Chapter 4 data and then uses tmap to do this:

```
load("ch4_db.RData")

lsoa_sf %>% left_join(lsoa_result) %>%
  tm_shape() + tm_fill("cost_pp", style = "kmeans",
                       palette = "YlOrRd", legend.hist = T) +
  tm_layout(frame = F, legend.outside = T, legend.hist.width = 1,
            legend.format = list(digits = 1),
            legend.outside.position = c("left", "top"),
            legend.text.size = 0.7, legend.title.size = 1)
```

However, this suffers from the invisibility problem described above. Mapping the rates using hexbins could overcome this. However, creating individual hexbins for each area (as was done in the Nottingham example in Figure 8.5) would still render the trends and spatial patterns difficult to interpret. A further option is to create a layer with a smaller number hexbins and then to summarise the lsoa_result data over those by allocating the LSOA prescription counts or costs (*not* the rates) to the hexbins in some way.

The code below creates a 10 km hexbin spatial layer:

```
hg = st_make_grid(lsoa_sf, 10000, what = "polygons", square = F)
hex = data.frame(HexID = 1:length(hg))
st_geometry(hex) = hg
```

There are further choices for the allocation:

1. They could be allocated to the hexbin that the LSOA centroid falls in (a point-in-polygon approach).

2. They could be allocated using a spatial interpolation approach, such as area-weighted interpolation.

Both of these involve additional but related assumptions. Using the LSOA geometric centroids assumes that the centroid is representative of the LSOA location. It is quite possible for some centroid locations to fall outside of the actual area they represent, especially for areas with convoluted shapes (i.e. when they do not approximate to a regular shape). Also, values may be allocated from parts of LSOAs that do not intersect (overlap) with the hexbin area. An area-weighted interpolation could overcome this problem as it allocates counts proportionate to the area of intersection between two spatial layers (such as the LSOAs and a hexbin layer).

The code snippets below illustrate these two approaches. First, a *point-in-polygon* approach. The first step is to convert the lsoa_result data table to a spatial points layer and then intersect this with the hexbin layer:

```
# link to the LSOA areas and convert to points
lsoa_sf %>% left_join(lsoa_result) %>% st_centroid() %>%
  # intersect with hexbin layer
  st_join(hex, join = st_within) -> ol
```

Then the intersection layer, ol, which now contains an attribute of the hexbin it intersects with (HexID), can be summarised over the hexbins. The code below defines does this and links back to hex to create hex_ssri:

```
st_drop_geometry(ol) %>% group_by(HexID) %>%
  summarise(hex_cost = sum(lsoa_tot_cost),
            hex_items = sum(lsoa_tot_items),
            hex_pop = sum(population)) %>%
  ungroup() %>%
  mutate(cost_pp = hex_cost/hex_pop,
         items_pp = hex_items/hex_pop) %>%
  right_join(hex) -> hex_ssri
# add the geometry back
st_geometry(hex_ssri) = hg
```

The populated hexagonal layer can be mapped as in Figure 8.8, with very different distributions for SSRI prescribing cost and counts:

```
tm_shape(hex_ssri)+
  tm_polygons(c("cost_pp", "items_pp"),
              palette = "viridis", style = "kmeans")
```

We now turn to *area-weighted interpolation*, which is done with the sf function st_interpolate_aw. This takes two polygon layers as inputs and has functionality that supports the summing of count variables, such as the cost, n and population variables over the target areas. This generates mean values for variables that are spatially intensive, like population density; it does not calculate weighted means. To overcome this, the data should be manipulated such that the

rates are converted to counts by population. This would support the calculation
of rates after conversion to the hexagon structure:

```
lsoa_sf %>% left_join(lsoa_result) %>%
    select(lsoa_tot_cost, lsoa_tot_items, population) %>%
    st_make_valid() %>% st_interpolate_aw(hex, extensive = T) %>%
    mutate(cost_pp = lsoa_tot_cost/population,
            items_pp = lsoa_tot_items/population) -> hex_ssri2
```

Figure 8.8 A map of SSRI antidepressant prescribing rates and costs in each hexbin grid
using a point-in-polygon approach

The rates can be calculated from the interpolated count data and mapped as in
Figure 8.9:

```
# map
tm_shape(hex_ssri2)+
    tm_polygons(c("cost_pp", "items_pp"),
                palette = "viridis", style = "kmeans")
```

The two approaches show the same broad pattern, with differences between the
binned results in Figures 8.8 and 8.9 being mainly in the rural and coastal fringes
(e.g. on the East and North East coast). This shows the difference between allocation based on LSOA geometric centroids and proportional areal overlap of the
LSOAs with the hexbins.

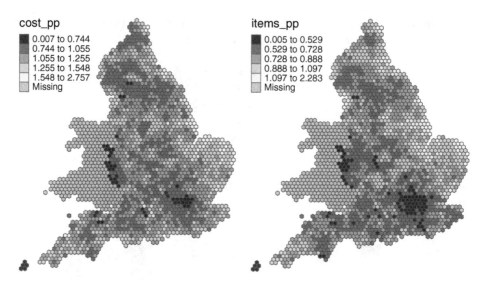

Figure 8.9   A map of rate of SSRI antidepressant prescribing rates in each hexbin grid using an areal weighting interpolation approach

8.6.4 Considerations

Prescribing data were selected, summarised, allocated to LSOA areas and then binned over summary areas (hexagonal areas) to examine the spatial variation in rates of pre-scribing. It is instructive to examine the results of the two binning datasets over hexbins generally and specifically using point-in-polygon approaches (Figure 8.8) and using areal interpolation (Figure 8.9), in light of their different assumptions.

In *general*, the binning process, whether using hexagons, square tiles or other shapes such as circles, provides a method for summarising data and relationships between variables or attributes over new geographical areas.

The first assumption is that a grid of 10 km is an appropriate scale over which to summarise data. This is a key consideration if these binned data are to be used for subsequent statistical analysis, because of the modifiable areal unit problem (MAUP) or ecological fallacy (Robinson, 1950; Openshaw, 1984a, 1984b). The basic idea is that if data are aggregated over different units then the observed patterns and relationship between variables will change with the changes in scaling, some-times in unexpected ways. The observed patterns may well change if a different scaling was used. You might want to examine the impacts on the distributions in Figures 8.7 and 8.8 with, say, a 5 km grid and a 20 km grid. Choices about the appropriate scale of observation should be made with an understanding of the scales over which the processes being examined operate.

Assumption 1: A 10 km hexagonal grid is an appropriate scale over which to aggregate LSOA prescribing counts.

It is also important to examine all of the *specific* assumptions behind the different steps in the process.

Second, the prescribing data were extracted from a database using a series of nested queries that selected, linked and manipulated data held in different tables in the database. The database has some 120 million records of individual prescriptions, for 2018, from which records related to SSRIs were extracted (or filtered, in dyplr terminology) and then summarised over individual GP practices to generate counts of the total cost and total number for each practice. Data on the number of patients registered at each GP practice in each LSOA were used to determine the proportion of patients in each GP practice, in each LSOA. These proportions were then used to generate counts of prescriptions in each LSOA and prescribing rates (cost per patient and number per patient).

Assumption 2: SSRI prescriptions were equally likely to be given to all patients.

Then the LSOA prescribing data were interpolated over hexagonal areas in two different ways. There are many different approaches for areal interpolation (see Comber and Zeng, 2019, for a review with illustrations coded in R) and each has its own assumptions.

The first approach used a point-in-polygon approach to implement an area-to-point interpolation (Martin, 1989). This allocated the LSOA *source zone* data to the hexagonal *target zones* using the spatial intersection of the LSOA centroids (geometric centre) with the target zones: the LSOA allocated to the target zone in which its centroid sits. This assumes that the centroid is representative of the population in that area. However, in reality the LSOA may have areas that are more densely populated than others and actually a more representative point location could be closer to more heavily populated areas: the centroid may not be representative of where the people live, and some kind of population-weighted centroids may be appropriate (Martin, 1996).

Assumption 3: The LSOA population centroids are representative of the mean location of patients in point-in-polygon or area-to-point approaches.

The second approach used an areal interpolation approach. This allocates the source zone attributes proportionately to the target zones based on the area of their intersection (Goodchild and Lam, 1980). This assumes that the relationship between the source zone attribute and the target zone area is spatially homogeneous. This may rarely be true in practice, but in the absence of other information, it remains a reasonable solution (Xian, 1995).

Assumption 4: In area weighting approaches, the relationship between the LSOA attribute and the hexbins is spatially homogeneous.

These assumptions are evident in the results which show subtly different distributions of the allocated values: area weighting allocates values to *all* of the hexagonal areas that areally intersect with the LSOAs, while point-in-polygon does not. This issue is more acute in the more rural regions that have larger LSOAs (recall LSOAs were designed to have ~1500 people), where the point-in-polygon results show more hexbins with a value of 0.

Key Points

- The use of binning was extended to a case study using the prescribing database introduced in Chapter 4.

- There are some inherent assumptions with binning approaches: the appropriateness of the bin size, and that prescribing relates to all patients in each input area equally.

- The prescribing area summaries (for LSOAs) were interpolated using two different approaches: by allocating the LSOA data to the hexbin in which the LSOA centroid falls (a point-in-polygon approach) and by applying a spatial interpolation approach (area-weighted interpolation).

- Each of these has embedded assumptions that need to be considered when the results are evaluated:

  - the centroid approach assumes that the centroid is representative of the LSOA location;

  - areal weighting assumes that the LSOA prescribing rate is evenly distributed across the LSOA.

8.7 SUMMARY

This chapter has introduced some advanced techniques for spatial data visualisation. It demonstrated some of the techniques for improving the visualisation of relationships and processes in the data, including cartograms, tiling and binning (pooling). A number of ways of doing these were illustrated and their advantages and limitations were described. The aim of the binning was to allow visual exploration of the data, but this process involves a number of different assumptions. The binned data *could* be used for analysis *but* this is likely to be severely affected by MAUP/ecological fallacy (Robinson, 1950; Openshaw, 1984a) issues unless it is very carefully done. An exploration of spatial trends in antidepressant prescribing in 2018 with socio-economic factors (exploring the 'geography of misery') was used to illustrate how advanced visualisation approaches can be applied in practice.

REFERENCES

Comber, A., Brunsdon, C., Charlton, M. and Cromby, J. (2020) The changing geography of clinical misery in England: Lessons in spatio-temporal data analysis. In J. Corcoran, M. Birkin and G. P. Clarke (eds), *Big Data Applications in Geography and Planning: An Essential Companion*. Cheltenham: Edward Elgar.

Comber, A. and Zeng, W. (2019) Spatial interpolation using areal features: A review of methods and opportunities using new forms of data with coded illustrations. *Geography Compass*, 13(10), https://doi.org/10.1111/gec3.12465.

Dorling, D. (1996) *Area Cartograms: Their Use and Creation*, Catmog 59. Norwich: School of Environmental Sciences, University of East Anglia.

Dougenik, J. A., Chrisman, N. R. and Niemeyer, D. R. (1985) An algorithm to construct continuous area cartograms. *Professional Geographer*, 37(1), 75–81.

Gastner, M. T. and Newman, M. E. J. (2004) Diffusion-based method for producing density-equalizing maps. *Proceedings of the National Academy of Sciences*, 101(20), 7499–7504.

Goodchild, M. F. and Lam, N. S.-N. (1980) *Areal Interpolation: A Variant of the Traditional Spatial Problem*. London, ON: Department of Geography, University of Western Ontario.

Harris, R., Charlton, M. and Brunsdon, C. (2018) Mapping the changing residential geography of White British secondary school children in England using visually balanced cartograms and hexograms. *Journal of Maps*, 14(1), 65–72.

Harris, R., Charlton, M., Brunsdon, C. and Manley, D. (2017a) Balancing visibility and distortion: Remapping the results of the 2015 UK General Election. *Environment and Planning A*, 49(9), 1945–1947.

Harris, R., Charlton, M., Brunsdon, C. and Manley, D. (2017b) Tackling the curse of cartograms: Addressing misrepresentation due to invisibility and to distortion. In *Proceedings of the 25th GIS Research UK Conference, Manchester*. Retrieved from http://huckg.is/gisruk2017/GISRUK_2017_paper_29.pdf.

Kuhn, H. W. (1955) The Hungarian method for the assignment problem. *Naval Research Logistics Quarterly*, 2(1–2), 83–97.

Martin, D. (1989) Mapping population data from zone centroid locations. *Transactions of the Institute of British Geographers*, 14(1), 90–97.

Martin, D. (1996) An assessment of surface and zonal models of population. *International Journal of Geographical Information Systems*, 10(8), 973–989.

Martin, D. (1997) From enumeration districts to output areas: Experiments in the automated creation of a census output geography. *Population Trends*, 88, 36–42.

Martin, D. (1998) 2001 Census output areas: From concept to prototype. *Population Trends*, 94, 19–24.

Martin, D. (2000) Towards the geographies of the 2001 UK Census of Population. *Transactions of the Institute of British Geographers*, 25(3), 321–332.

Martin, D. (2002) Geography for the 2001 Census in England and Wales. *Population Trends*, 108, 7–15.

Nusrat, S. and Kobourov, S. (2016) The state of the art in cartograms. *Computer Graphics Forum*, 35(3), 619–642.

Olson, J. M. (1976) Noncontiguous area cartograms. *Professional Geographer*, 28(4), 371–380.

Openshaw, S. (1984a) Ecological fallacies and the analysis of areal census data. *Environment and Planning A*, 16(1), 17–31.

Openshaw, S. (1984b) *The Modifiable Areal Unit Problem*, Catmog 38. Norwich: Geo Abstracts.

Robinson, W. (1950) Ecological correlations and the behavior of individuals. *American Sociological Review*, 15, 351–357.

Tobler, W. (2004) Thirty five years of computer cartograms. *Annals of the Association of American Geographers*, 94(1), 58–73.

Xie, Y. (1995) The overlaid network algorithms for areal interpolation problem. *Computers, Environment and Urban Systems*, 19(4), 287–306.

(9)

EPILOGUE ON THE PRINCIPLES OF SPATIAL DATA ANALYTICS

9.1 WHAT WE HAVE DONE

In this first section of this final chapter, we attempt to bring together the areas we have covered in this book into an overarching framework – and to some extent present our own justification of why we have presented things in the way that we have. The subsections below set out a number of approaches we have taken – essentially a list of *things that we have done* in putting together this book.

9.1.1 Use the tidyverse

Much of the R code in this book makes use of the *tidyverse* – essentially a set of R libraries intended to promote a 'tidy data' approach to data manipulation, modelling and graphics. A key set of characteristics of tidy data are data frames satisfying the following requirements:

- Each row corresponds to one observation unit.
- Each column corresponds to one variable.
- Each value corresponds to one cell.

In particular, this attempts to avoid situations where a value masquerades as a variable – such as a cross-tabulation where rows are locations, and columns are months for average rainfall. In that instance, the tidy alternative would be to regard a row as a rainfall measure, with columns for average rainfall, location and month. This format resembles the structure of a relational database – and makes it easier, for example, to join other data tables with details indexed by location or by month. Similarly, the format also avoids situations where a single cell in the table stores multiple information (e.g. the type of a house – terraced, detached, etc.) and the number of bedrooms (e.g. 'T2' for a two-bedroomed terraced house).

Invariably, the two distinct parts of the value will be analysed in their own right, and the need to code to separate them is unhelpful.

Many of the tools in the dplyr package – part of the tidyverse – can be divided into two broad groups:

- Conversion to and from tidy format

- Manipulation in tidy format.

The first is useful, since much data acquired do not fit into this structure, and so converting *to* tidy format allows further manipulation to take place. Manipulation can involve operations such as group-wise summaries, subset selection, data transformation and ordering. Finally, the ability to convert *from* tidy format to others is also useful – either because a further user may require the data in a different format or because although a non-tidy format (such as the cross-tabulated rainfall example) may not be a very good way to *manipulate* data, it may make a more readable table. Tidy data are intended for effective and unambiguous data manipulation, rather than effective human readability – and for this reason, the results of tidy data analysis may (and sometimes should) be recast into alternative data formats.

A typical example is shown in Figure 9.1. Here data are converted into a tidy format, a subset of the data is selected, this subset is transformed. and finally, an analysis, a visualisation and a recasting into another format are carried out in parallel to the resultant tidy dataset. The tidy operations are shown in green, the others in brown.

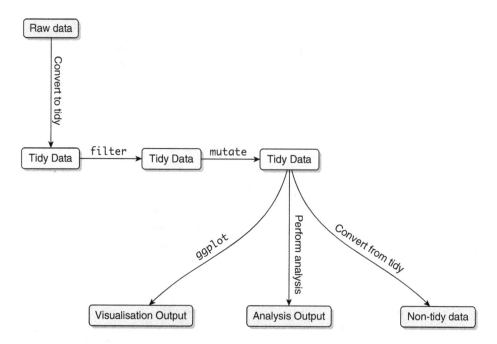

Figure 9.1   A possible tidyverse data analysis

Where explicit, a `dplyr` function is given to demonstrate the appropriate R procedure to carry out the operation to the data. Thus `filter` and `mutate` are useful tools for data transformation and selection. Also the `ggplot` function is well suited to creating graphical representations of tidy data.

9.1.2 Link analytical software to databases

The approach in the previous subsection essentially encourages 'data-frame-oriented' thinking – this is important when working directly with databases dealing with data not stored directly in R's allocated memory. Such approaches are important when the size of the datasets exceeds the memory capacity of R – in this case data are stored either on a local hard disk, or possibly in the cloud. `SQLite` is a database that is relatively easy to install (unlike many others it does not require a distinct 'server' process to run). Databases such as this understand commands written in the *Structured Query Language* (SQL). Typically R 'talks' to the database by sending requests written in SQL and in return receives either a selected subset of the database or the result of some operation applied the database. Ideally, if the range of operations that SQL offered was as flexible as that provided by R then we could work almost exclusively in SQL, using R for graphics and data summaries. However, this is not actually the case: SQL generally has a much more limited set of mathematical functions – for example, it is not possible in `SQLite` to compute the median of a variable or to compute its square root directly. In practice, currently one has to select subsets from `SQLite` into R and process them in R. If a key aim is data reduction, this is fine – for example, to compute countywise median values for data in which one column specifies county, the data can be selected into R in 'chunks' of counties. However, at the moment the implication of this is that R and the SQL database have to work in a closely coupled (Goodchild et al., 1992) sense. One helpful tool is `dbplyr` which 'cocoons' the database so that it appears as a standard tidyverse tibble as illustrated in Figure 9.2: the solid lines are the

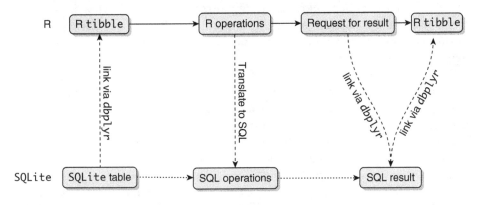

Figure 9.2 R and SQLite close coupling

user-visible workflow in R, the dotted lines are the 'behind the scenes' SQL work-flow and the dashed lines show the coupling. Clearly, this is very close coupling indeed. In this way manipulation of the SQL database is carried out in a way that is syntactically similar to working with tidyverse operations in R – although limita-tions in the function availability on the database side may imply that some opera-tions that would be valid with internally stored tables would not work in this situation. However, it is important to note that when working with very large data, such coupling between R and external databases is now essential if some reason-ably sophisticated data analysis is needed. Note that the 'request for result' is often implicit – for example, the assignment of a dplyr pipeline to a new tibble.

9.1.3 Look through a spatial lens

A key aim of this book is not just to consider data science, big data and the linkage of R with external databases, but also to consider the manipulation and analysis of spatial data within this framework. As things stand, one key development here is the use of *simple features* to represent spatial information. To the user in R, geo-graphical objects appear in a tibble with a column called geometry. Each entry in this column is a kind of pseudo-value, containing geometrical information about an individual geographical entity – such as a point (e.g. location of a crime), a polygon (e.g. a county) or a line (e.g. a road). The other columns on the same row contain non-geometrical information about the geographical entity (e.g. the name of a county, or its current population).

The simple feature format is an internal structure used by R that may also be used in relational databases. In addition to providing a way of *storing* geograph-ical information, the sf package (Pebesma et al., 2019) in R also offers a number of processing tools – essentially methods of processing the information in the geometry column (e.g. computing buffers or identifying, for a pair of tables, which individual geometry objects intersect). sf works well in conjunction with tidyverse operations, and tibbles with geometry can be treated in similar ways to other tibbles (e.g. group followed by summarise applied to an sf will merge the geometries of the summarised distinct groups). Also, an st_join (from sf) performs a spatial join between two sf files, where records with intersecting geometries are joined together.

Note that although the above provides a way of manipulating spatial data in R, this will by default work on data *in memory* and so will not work well with very large datasets, where an external database would be more appropriate. Extensions to data-base servers, such as SpatiaLite (Furieri, 2019) – a spatial extension of SQLite – are helpful here. However, at this stage the degree of coupling between sf and SpatiaLite is not strong – a diagram such as Figure 9.2 does not apply here. As a result, it is often more practical to just 'chunk' spatial data into pieces that may be processed in R or to process spatial data externally (e.g. in SpatiaLite), save the data into a known file format (e.g. a shapefile) and then read the processed data into R.

9.1.4 Consider visual aspects

Another important aspect of data analysis is visualisation – it is quite often possible to see overall trends in the data or quickly identify unusual or counter-intuitive observations (or groups of observations) if the data are represented in image form. This is a general issue for data analysis and data science. There are also particular topics in visualisation relating to spatial information.

We consider two key visualisation issues here. The first is a more formal framework for visualisation using the structures suggested by Wilkinson (2012) and implemented in R via the ggplot2 package (Wickham, 2016). This provides a way of specifying visualisations by linking variables in a tidy data frame to *aesthetics* such as colour, *x*-position, *y*-position and size, and then specifying the *geometry*, details of the form of the visualisation (e.g. a point plot, a line plot) and so on. Again, this is appropriate here as it meshes well with the tidy data model intro-duced in other parts of this book.

From a spatial perspective, ggplot2 is also helpful, since it offers a geometry (geom_sf) that works with simple features objects, and draws maps. Another option for spatial data is the tmap package (Tennekes, 2018), which focuses on visualising spatial data and uses similar ideas of aesthetics to map variables to visual characteristics, but also enables interactive, zoomable and panable maps to be created with backdrops from third parties, such as OpenStreetMap.

The second key issue we focus on is that of visualising larger datasets. Some forms of visualisation (e.g. scatter plots) do not 'scale up' well. The forms present-ing the most problems are often those in which there is a one-to-one correspond-ence between observations in the data table and objects in the visualisation. This is the case with scatter plots – each plotted point is associated with a pair of variables in a row in the data frame. This generates two difficulties if the dataset is large:

- The storage overhead for the graphic becomes large if it is stored in a vector-based format such as PDF. This also suggests that the graphics take a long time to draw.

- Pattern visibility is poor in areas of the plot with a large number of points. Some areas may become completely covered in points several times over, and relative density within these areas cannot be perceived.

The second problem can be overcome by using points with some degree of transparency – but this does not address the first problem. Generally an approach involving *graphical reduction* should be used here – that is, summarising the 'raw' graphical objects by a smaller number of objects that still convey features seen in the original information. One approach to this is *binning* – dividing the range of the raw data into a number of intervals or bins (or two-dimensional tessellations) and assigning each original item to one of these. Information in each bin is then sum-marised, and the visualisation produced uses these bins as the unit of geometry, with aesthetics based on the summarised information – an example overlaying the raw scatter plot on hexagonal binning is given in Figure 9.3. Since the number of bins should be notably less than the original number of observations, the problems listed above should be overcome.

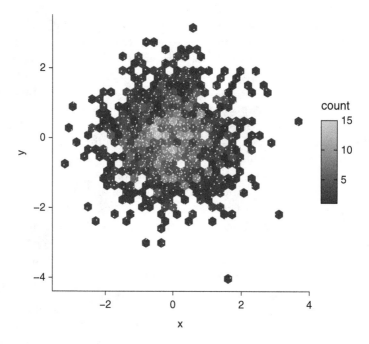

Figure 9.3   Example of hexagonal binning

An alternative approach to this is *smoothing*, where a moving-window focal summary statistic is used, rather than a division of the range of a variable into intervals. For a number of focal points along the range (or a regular lattice filling it in the two-dimensional case) observations falling in a window centred on the points are summarised. The summaries are often weighted so that observations nearer the focus have more influence. The geometric units are then the points, with the aesthetic linking to the value of their summary statistics. An example is given in Figure 9.4, where a window-based density estimate is used to provide contoured density estimates for the same data as Figure 9.3. We cover both of these approaches, again looking at how they may be used in a tidy data framework, via ggplot2.

9.1.5 Consider inferential aspects

The above subsections give a good idea of the practice of manipulating and visualising spatial and other data. It is often necessary to analyse data in other ways than visual. Two common activities are attempting to understand the processes giving rise to the data, and predicting future values of the data, or current values where no observation is available. The methods are considered in depth in other areas, but emphasis here is on gaining some insight into process, even in a predictive scenario. We would even take the step of strongly advising **against** a black box prediction approach. The term 'big data hubris' is used by Lazer et al. (2014) to indicate an 'often implicit assumption that big data are a substitute for, rather than a supplement to, traditional data collection and analysis'. A number of cautionary tales exist – one of the best known is that of Google Flu Trends (GFT).

After a successful early period, this attempt to predict the incidence of influenza began to perform badly. The methodology used was primarily guided by considerations of what Google search terms made good predictors, based on a set of around 50,000 terms. However, certain aspects of the process were not included in the model, such as seasonality. Some of the search terms were essentially proxies for time of year, such as searches relating to high school basketball. As Lazer et al. (2014) put it: 'In short, the initial version of GFT was part flu detector, part winter detector.' In addition, Google itself altered its search algorithms and features (e.g. adding a *suggested search* feature), altering the way people used Google to search the internet, and potentially altering the process being analysed here.

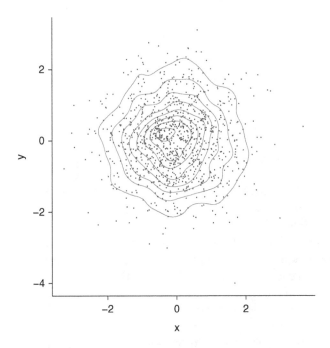

Figure 9.4   Example of density smoothing

Here we argue that although analysis of very large datasets (big data) is a useful tool, it should not be regarded as 'the end of theory' (Anderson, 2008). Failure to consider underlying processes and working with a 'prediction is everything' paradigm can lead to several real-world problems. Thus, among other things, we still argue for the use of existing tools of statistical inference (essentially to evaluate hypotheses or assess the accuracy of model calibration) in a data science environment. There are times when the largeness of a dataset brings on its own difficulties, particularly for hypothesis testing – see, for example, Nester (1996).

9.2 WHAT WE HAVE FAILED TO DO

This section is essentially a review of things we have not covered but do think are important. They may have been omitted for a number of reasons – three prominent ones are that:

- although important, the book has constraints on its practical size;
- although important, they did not fit in with the structure of this particular book;
- they are very well covered elsewhere.

Many omissions can be justified by some combination of the above. However, in the closing sections of the book, we feel it would be helpful to draw some attention to these ideas, and where possible to suggest pointers to further reading. This section, like the previous one, can be thought of as a list, but this time of *things that we have not done*.

9.2.1 Look at spatio-temporal processes

One of the key promises of ever larger datasets (with data science) is that they may be delivered in close to real time. This suggests that information about processes will not only be spatially tagged but also time-stamped. Thus, large databases recording processes evolving in time and space are becoming commonplace. As with spatial information, spatio-temporal information also has its own research questions and issues – spatio-temporal processes are different. The differences reflect many aspects covered in this book – in particular, storage, manipulation, visualisation and analysis. Taking a cue from Tobler (2004), we offer that in an age where many data are spatio-temporal – they are collected some*where* and some*time* – there is a need to consider the process under investigation and critically consider its periodicity or temporal association.

One interesting aspect is that in the same way that aggregated spatial data have a *modifiable areal unit problem* (MAUP; Openshaw, 1984), a similar problem exists for temporal data. This is considered, for example, by Stehle and Kitchin (2019). This can have repercussions for the way temporal information is stored.

Furthermore, one can couple the temporal problem here with the MAUP for data that are both spatially and temporally tagged. Essentially we have a *modifiable areal/temporal unit problem*.

In addition, temporal processes can exhibit both ongoing trends (e.g. global warming) and cyclic effects (e.g. seasonal patterns in rainfall). Both kinds of pattern can interact with space – summer in the northern hemisphere is winter in the southern, and temperature increase due to global warming will not take the same value everywhere. A number of modelling techniques exist to address spatio-temporal processes – for example, the *space time autoregressive integrated moving average* (STARIMA) model (Pfeifer and Deutsch, 1980) builds on existing stochastic process modelling approaches to consider processes evolving in space and time. An example of predicting traffic flows is provided by Duan et al. (2016). However, as yet there are no standard methodologies for dealing with such models in a tidy data framework.

A further issue exists with cyclic data – such as time of day – where standard summary statistics are unhelpful (Brunsdon and Charlton, 2006). For example, representing time of day on a linear scale using decimal hours, ranging from 0 to 24, and looking at a set of observed events occurring just before and just after midnight would yield an average of around 12 – representing noon. However, this is quite clearly misleading. In fact there is a need to redefine averages, variance measures and other statistics for cyclic data – see, for example, Fisher (1993). Indeed, there exist probability distributions, regression models and visualisation techniques that are unique to cyclic variables. Couple these with spatial dependence and some quite complex models emerge. Similarly, visualisation techniques for spatial cyclic data also require specific attention (Brunsdon et al., 2007).

Thus, spatio-temporal data science is a complex and wide-ranging topic – certainly one that is important, but perhaps needing an entire book to deal with in a useful and rigorous manner.

9.2.2 Look at textual data

Another characteristic of contemporary data is that much of them are textual. Although character strings exist in cells of data frames, generally they tend to be used as factor variables (i.e. they are designed to take a small number of fixed values, such as the colour of an object, or the political party returned in an election), or an identifier (e.g. somebody's name) or as a way of making an informal note relating to an observation. The kinds of data referred to here relate to the content of tweets, Facebook entries and so on. Here the focus of the data is full textual content – complete phrases, or possibly sentences or paragraphs. Identifying patterns in the content of this kind of textual information is a specialised analysis task. Interestingly, it is possible to represent and analyse large textual databases in a tidy data format – a useful text here is Silge and Robinson (2017).

9.2.3 Look at raster data

The simple features spatial data model as implemented in R is essentially designed to work with spatial objects in a vector framework. Although rasters could be envisioned in this framework they would essentially be either a set of point features arranged on a rectangular grid, or a lattice of rectangles. Thus, much of the way R deals with rasters in a tidy data way is just to convert raster data into a representation of this kind, and then treat it as a vector-based spatial object. However, the Raster package in R provides an alternative system, where operations are carried out in an entirely raster-based environment, which is generally much faster. As with spatio-temporal data, this is an important issue, but perhaps one deserving of a book in itself.

9.2.4 Be uncritical

In addition to considering techniques of data analysis, many argue (such as Kitchin, 2014) for critical reflection on the assumptions often adopted by the media and technical literature, in particular challenging the notion of big data and associated data science as objective and all-encapsulating (Iliadis and Russo, 2016). This has led to the field of *critical data studies* (CDS). Kitchin and Lauriault (2014) propose that CDS should consider 'the technological, political, social and economic apparatuses and elements that [constitute and frame] the generation, circulation and deployment of data'. Iliadis and Russo (2016) argue that CDS should ultimately encourage us to 'think about big data science in terms of the common good and social contexts'. We consider that an important goal here is to encourage the practitioners of data science to do this. We would also extend this remit to consider critical views on some of the more technical aspects of the data analysis. Quoting O'Neil and Schutt (2014): 'We'd like to encourage the next-gen data scientists to become problem solvers and question askers, to think deeply about appropriate design and process, and to use data responsibly and make the world better, not worse.'

Taking on board Derman's (2011) *Hippocratic Oath of Modeling*, in particular the declaration that 'I understand that my work may have enormous effects on society and the economy, many of them beyond my comprehension', one could easily add 'the environment' to 'society and the economy'. This sits uneasily with Cukier and Mayer-Schoenberger's (2013, p. 40) outlining of the characteristics of big data – although they do argue also that 'we must adopt this technology with an appreciation not just of its power but also of its limitations'. A typical concern may then be the wider implications of misinterpreting the inferential aspects of a particular analysis. As O'Neil and Schutt (2014, p. 354) warn: 'Even if you are honestly skeptical of your model, there is always the chance that it will be used the wrong way in spite of your warnings.' Although the ideas here are not the key focus of this book, we would strongly urge data science practitioners to be aware of the debates here, and follow the advice offered.

9.3 A SERIES OF CONSUMMATIONS DEVOUTLY TO BE WISHED[1]

Having reflected on the content of this book, and other things not covered in this book that we feel a spatial data scientist should be familiar with, we now move on to consider a list of potential advancements in data science that we feel would be beneficial. These are not necessarily predictions – although as they are things that we would like to see we hope that they *do* turn out to be. In many ways they are based on the 'holes' in the tools, capabilities and concepts of spatial data science that we feel we have noted in considering the previous two sections of this chapter.

9.3.1 A more integrated spatial database to work with R

This is a practical desire. Earlier it was noted that although there are spatially aware versions of database servers (e.g. `SpatiaLite`) there is not the close coupling between R and `sf` that there is between non-spatial databases and R, via `dplyr`. As a means of working with spatial data it would be useful if such a coupling could be made – possibly as an extension of `sf`.

9.3.2 Cloud-based R computing

This is something that is perhaps already beginning to occur, for more demanding R-based tasks. In particular, a number of tasks in R may be carried out in parallel – a useful technique for speeding up some kinds of code – and although exploiting multicore machines is commonplace, facilities to share tasks across networks of computers on the cloud could have notable potential.

9.3.3 Greater critical evaluation of data science projects

This was mentioned in the previous section. We argue here that a stronger critical awareness of work being done with data, and in a data science framework, should be viewed not only as a technical exercise, but also as an action that may influence policy and have social, economical and environmental consequences. Whereas this is not something that can be achieved by the provision of software or hardware, we hope that discussion of such topics in the academic literature, at conferences (both academic and professional) and in courses and degree programmes in data-science-related topics will lead to greater consideration of these matters.

---

[1]*Hamlet*, via Yvette Livesey and Anthony H. Wilson's 'Dreaming of Pennine Lancashire' (https://www.lancashiretelegraph.co.uk/news/1015236.our-dream-to-improve-region-for-everyone/). OK, it's a shopping list really.

REFERENCES

Anderson, C. (2008) The end of theory: The data deluge makes the scientific method obsolete. *Wired*, 23 June. www.wired.com/science/discoveries/magazine/16-07/pb_theory.

Brunsdon, C. and Charlton, M. (2006) Local trend statistics for directional data – A moving window approach. *Computers, Environment and Urban Systems*, 30(2), 130–142.

Brunsdon, C., Corcoran, J. and Higgs, G. (2007) Visualising space and time in crime patterns: A comparison of methods. *Computers, Environment and Urban Systems*, 31(1), 52–75.

Cukier, K. and Mayer-Schoenberger, V. (2013) The rise of big data: How it's changing the way we think about the world. *Foreign Affairs*, 92(3), 28–40.

Derman, E. (2011) *Models Behaving Badly: Why Confusing Illusion with Reality Can Lead to Disaster, on Wall Street and in Life*. Chichester: John Wiley & Sons.

Duan, P., Mao, G., Zhang, C. and Wang, S. (2016) STARIMA-based traffic prediction with time-varying lags. In *2016 IEEE 19th International Conference on Intelligent Transportation Systems (ITSC)*. Piscataway, NJ: IEEE.

Fisher, N. I. (1993) *Statistical Analysis of Circular Data*. Cambridge: Cambridge University Press.

Furieri, A. (2019) *SpatiaLite – Spatial Extensions for SQLite*. Gaia-SINS. www.gaia-gis.it/spatialite-2.1/SpatiaLite-manual.html.

Goodchild, M., Haining, R., Wise, S. et al. (1992) Integrating GIS and spatial data analysis: Problems and possibilities. *International Journal of Geographical Information Systems*, 6(5), 407–423.

Iliadis, A. and Russo, F. (2016) Critical data studies: An introduction. *Big Data & Society*, 3(2), 1–7.

Kitchin, R. (2014) Big data and human geography: Opportunities, challenges and risks. *Dialogues in Human Geography*, 3(3), 262–267.

Kitchin, R. and Lauriault, T. (2014) Towards critical data studies: Charting and unpacking data assemblages and their work. The Programmable City Working Paper 2, Preprint. https://ssrn.com/abstract=2474112.

Lazer, D., Kennedy, R., King, G. and Vespignani, A. (2014) The parable of Google Flu: Traps in big data analysis. *Science*, 343(6176), 1203–1205.

Nester, M. R. (1996) An applied statistician's creed. *Applied Statistics*, 45(4), 401–410.

O'Neil, C. and Schutt, R. (2014) *Doing Data Science: Straight Talk from the Front Line*. Sebastopol, CA: O'Reilly.

Openshaw, S. (1984b) *The Modifiable Areal Unit Problem*, Catmog 38. Norwich: Geo Abstracts.

Pebesma, E., Bivand, R., Racine, E., Sumner, M., Cook, I., Keitt, T. et al. (2019) Simple features for R. https://cran.r-project.org/web/packages/sf/vignettes/sf1.html.

Pfeifer, P. E. and Deutsch, S. J. (1980) A three-stage iterative procedure for space-time modeling. *Technometrics*, 22(1), 35–47.

Silge, J. and Robinson, D. (2017) *Text Mining with R: A Tidy Approach*. Sebastopol, CA: O'Reilly Media.

Stehle, S. and Kitchin, R. (2019) Real-time and archival data visualisation techniques in city dashboards. *International Journal of Geographical Information Science*, 34(2), 344–366.

Tennekes, M. (2018) tmap: Thematic maps in R. *Journal of Statistical Software*, 84(6), 1–39.

Tobler, W. (2004) On the first law of geography: A reply. *Annals of the Association of American Geographers*, 94(2), 304–310.

Wickham, H. (2016) *ggplot2: Elegant Graphics for Data Analysis*. Cham: Springer.

Wilkinson, L. (2012) The grammar of graphics. In J. E. Gentle, W. Härdle and Y. Mori (eds), *Handbook of Computational Statistics* (pp. 375–414). Berlin: Springer.

INDEX

CPSIA information can be obtained
at www.ICGtesting.com
Printed in the USA
BVHW062037160622
639812BV00004B/63